Superstring Theory: The DNA of Reality
Part II

Professor S. James Gates, Jr.

THE TEACHING COMPANY ®

PUBLISHED BY:

THE TEACHING COMPANY
4151 Lafayette Center Drive, Suite 100
Chantilly, Virginia 20151-1232
1-800-TEACH-12
Fax—703-378-3819
www.teach12.com

Copyright © The Teaching Company, 2006

Printed in the United States of America

This book is in copyright. All rights reserved.

Without limiting the rights under copyright reserved above,
no part of this publication may be reproduced, stored in
or introduced into a retrieval system, or transmitted,
in any form, or by any means
(electronic, mechanical, photocopying, recording, or otherwise),
without the prior written permission of
The Teaching Company.

ISBN 1-59803-162-7

S. James Gates, Jr., Ph.D.

John S. Toll Professor of Physics, University of Maryland at College Park

S. James Gates completed his undergraduate education and received two B.Sc. degrees (in mathematics and physics) at the Massachusetts Institute of Technology. His Ph.D. (in physics) was conferred for studies of elementary particle physics and quantum field theory. His Ph.D. thesis on *supersymmetry* was the first devoted to this subject at MIT. Dr. Gates's postgraduate studies started as a Junior Fellow of the Harvard Society of Fellows and ended with an appointment at Caltech. Faculty appointments began at MIT and later continued at the University of Maryland at College Park (1984–present). From 1991–1993, Dr. Gates served as physics professor and departmental chair at Howard University. In July 1998, he was named the first John S. Toll Professor of Physics, becoming the first African-American to hold an endowed chair in physics at a major research university in the United States. The Washington Academy of Sciences named him its 1999 College Science Teacher of the Year.

Professor Gates has authored or co-authored more than 180 research papers published in scientific journals, and one book, and contributed numerous articles to several books. His research, in the areas of the mathematical and theoretical physics of supersymmetric particles, fields, and strings, covers such topics as the physics of quarks, leptons, gravity, superstrings and heterotic strings, and unified field theories of the type first envisioned by Albert Einstein. Dr. Gates travels widely, speaking at national and international scientific meetings.

A member of the American Physical Society (APS), Sigma Xi, and the National Society of Black Physicists, Dr. Gates is also a past president of the NSBP and has served on the executive board of APS. He has served as a consultant for the National Science Foundation, Department of Energy, Department of Defense, Educational Testing Service, and Time-Life Books. He was the first recipient of the APS Bouchet Award and is a Fellow of the APS and NSBP; in 1997, he received the Martin L. King, Jr. Leadership Award from MIT. He is also a member of the 62nd College of Distinguished Lecturers of Sigma Xi and the board of directors of the Quality Education for Minorities Network (QEM).

The work of Dr. Gates and others was highlighted on "The Path of Most Resistance," in the PBS series *Breakthrough: The Changing Face of Science in America*. Dr. Gates has appeared in three additional PBS science documentaries, including *Einstein's Big Idea*. In March 1998, he appeared on the simultaneous C-Span television broadcast and Internet cybercast of the second Millennium Lecture by Professor Stephen Hawking from the East Room of the White House. Professor Gates was asked to provide comments on the topic of supersymmetry for broadcast and live audiences, including U.S. President William J. Clinton.

Table of Contents
Superstring Theory: The DNA of Reality
Part II

Professor Biography i
Course Scope 1

Lecture Thirteen	Gauge Theory—A Brief Return to the Real World 4
Lecture Fourteen	Princeton String Quartet Concerti—Part I 22
Lecture Fifteen	Princeton String Quartet Concerti—Part II 39
Lecture Sixteen	Extra Dimensions—Ether-like or Quark-like? 57
Lecture Seventeen	The Fundamental Forces Strung Out 76
Lecture Eighteen	Do-See-Do and Swing Your Superpartner—Part I 93
Lecture Nineteen	Do-See-Do and Swing Your Superpartner—Part II 110
Lecture Twenty	A Superpartner for Dr. Einstein's Graviton ... 127
Lecture Twenty-One	Can 4D Forces (without Gravity) Love Strings? 143
Lecture Twenty-Two	If You Knew SUSY 160
Lecture Twenty-Three	Can I Have That Extra Dimension in the Window? 177
Lecture Twenty-Four	Is String Theory the Theory of Our Universe? 194

Timeline 210
Glossary 217
Biographical Notes 237
Bibliography 245

Superstring Theory: The DNA of Reality

Scope:

This course aims to provide a non-technical and accessible description of the central foundational concepts and historical development of the topic in theoretical physics called *superstring/M-theory*. These lectures place this topic in the context of the more general development of mathematical and scientific thought that can be traced from the ancient realms of Egypt and Greece to medieval Iraq, Renaissance Europe, and the present.

By the end of the course, students will gain insight not only into the strange new world of superstring/M-theory but also into the central role of mathematics as the empowering element of human creativity driving the conception of science through theoretical physics. Although mathematics plays a central role in this story, it is kept to a minimum in the lectures. This is possible because of an almost unique capability of the courses produced by The Teaching Company.

For years, I have been asked to write a book covering this topic, largely because I have made more than 100 non-technical presentations on superstring theory since 1988. But no ordinary book would be capable of conveying to most people the mathematical ideas that provide the foundation of this topic. Unless and until it receives observational support, superstring/M-theory will be all about mathematics, but mathematics is largely inaccessible without highly specialized training. The experience of lecturing to nonscientists on this topic suggested to me that visual media, both still and animated, provide the key to solving this problem. The video format of Teaching Company courses is an exquisite platform for using computer-generated imagery to augment conventional lectures and books. Essentially, in these lectures, I use computers to "play" mathematics in much the same way that a musician plays scores.

The course begins with a description of this approach and a cursory look at the concept of the string. We'll explore the strange realization that understanding the universe at its largest scales requires knowledge of the smallest structures and their behavior, together with mathematics. We'll also look at the role of human creativity in the conception of science, and we'll discuss a number of not-so-well-known properties of mathematics as a tool for science.

Superstring/M-theory is sometimes presented as a radical break with all preexisting scientific thought. To counter this notion, we'll pay some attention to the known structure and rules of the universe at the very smallest scales. This discussion will establish many concepts that reappear in superstring/M-theory, in particular, the concept of the quantum world.

For many years, physicists largely ignored the fact that their accepted descriptions of the largest structures and the smallest structures in the universe were incompatible. We'll see how Stephen Hawking used black holes to force a crisis in theoretical physics. The only known way out of this crisis begins with *bosonic string theory*.

We'll also review Einstein's theory of special relativity and its role in string theory, noting the presence of time as the fourth dimension and the largely overlooked role of a structure that can be called *Einstein's Hypotenuse*. The dual requirements of quantum theory and the theory of special relativity in bosonic string theory lead to a description of a *tachyon*—a particle capable of destroying the known laws of physics.

Next, we turn to two little-known properties of the electron in the quantum world that, in effect, banish the tachyon and create a new generation of spinning strings and superstrings. We return briefly to the real world for a look at electricity and magnetism in preparation for a leap in the development of this subject to the *heterotic string*, conceived in 1984. This is the first mathematical construct that realized Einstein's long-sought dream of a *unified field theory*.

We next turn to the widely discussed possibility of hidden dimensions and the little-discussed alternative, followed by an exploration of the manner in which second-generation string theories describe all forces, including gravity. We'll pay some attention to the modifications implied by the newer strings, noting the rigorous mathematical and logical support that has been given to the conjecture of Stephen Hawking. We'll also discuss new forms of energy and matter, called *superpartners*, and learn about the superpartners for the particle zoo of the quantum world, including the superpartner for Einstein's graviton, which leads to *supergravity*.

Toward the end of the course, we'll look at current attempts to use concepts from string theory to gain increased understanding of the forces and structures of matter inside the proton and neutron. We'll explore a little-discussed "hidden dimension," radically different from all others, and its role in the concept of *superspace*. We'll close with the ultimate *supergravity theory*, associated with a world of eleven hidden dimensions, and its connection to the mother theory of strings, M-theory. As the lectures come to an end, we'll look toward the possibility of elucidating unsolved problems and meeting the challenges of this class of mathematical and scientific ideas.

Lecture Thirteen
Gauge Theory—A Brief Return to the Real World

Scope:

String theory is not yet science; it is mathematics. But the peculiar thing about the mathematics of string theory is that it looks like the mathematics we use to describe the real world. In some ways, the mathematics of string theory is inspired by known science, and it ought to begin to make predictions to be tested, but it is not at that point yet. In this lecture, the starting point is the year 1861, the year in which a young Scottish physicist, James Clerk Maxwell, first understood how we are able to see color. Maxwell was the physicist upon whose shoulders Einstein most clearly stood in producing the theory of special relativity. The ultimate unsuccessful quest of Maxwell and Einstein was the same: to find a unified field theory that would account for both the electromagnetic force and gravity.

Outline

I. A paper Maxwell wrote in 1873, entitled "A Treatise on Electricity and Magnetism," contains four equations which led to an understanding of light.

  A. Maxwell's equations imply that light is the propagation of an electromagnetic wave. These equations require photons to change as they move through both space and time in order to describe electric and magnetic fields of force.

  B. To describe light, these fields can be depicted as "playing a game of leapfrog." The electrical field changes in space and time and generates a magnetic field, which in turn, changes in space and time and generates an electrical field, and so on. The speed at which this occurs is the speed of light.

  C. This was Maxwell's great discovery, and all of our current communication technology can be traced back to these four equations.

  D. In 1888, Heinrich Hertz constructed an experimental device to detect Maxwell's waves of electromagnetic energy.

  E. Maxwell tried to describe gravity in mathematical terms 50 years before Einstein tried to modify his equations describing gravity to include electricity and magnetism. Both failed.

II. The concept of *change* of the photon is important to understand electricity and magnetism (also collectively called *electromagnetism*).

  A. Our familiar animation of the Pythagorean Theorem is set in motion to show changes in the areas of the blue square and the brown square.

  B. A second animation with two bowls of ping-pong balls, one fuller than the other, shows that even if two objects are different, the change in each can be the same.

  C. Maxwell's equation implies a similar relationship between electricity and magnetism, which can be measured in a laboratory, and photons, which are not directly measurable. It is the change in photons which produces the electric and magnetic fields.

  D. The same electromagnetic fields can be produced by different photons, each undergoing the same change. Since the fields are produced by changes in photons, not by the photons themselves, many photons are indistinguishable. This is an important idea in physics, called *gauge invariance*. Different photons that produce the same electromagnetic fields are related to each other by a quantity (*gauge parameter* or *Kemmer angle*) that is similar to an angle in the real world.

  E. By imagining a box filled with photons, we can consider the question of how much energy in the box is due to the photons.

   1. More than one photon can produce a specific electric or magnetic field. Thus, there could be two boxes in which the electric and magnetic fields were the same and which could possess the same electromagnetic energy, even though the photons in each box were different. The important point is that the changes in the photons could be the same.

   2. The energy in each box must depend on the photons in some way since, if there were no photons at all in the boxes, there would be no electromagnetic energy.

3. The fact that it is impossible to tell which photon produces how much energy leads to the idea of *charge conservation*, first understood by Emmy Noether. The *Noether theorem* implies that if there are changes in a quantity that is not directly measurable, but that some physically measurable quantities don't depend on those changes, then there must be conserved currents.
4. What is the conserved quantity associated with photons? The answer is electric charge. Due to Noether's theorem, it is understood that the conservation of electric charge is actually an implication about the ability of two (or more) distinct photons to produce the same electric or magnetic field.

III. In the quantum world, the laws of motion are very different from the laws of motion in our world. This has an important implication for charge.
   A. We consider a second animation which uses a bowl to compare motion in our world with that of denizens in the quantum world. In our world, balls set in the same position in the same way all follow the exact same path (a result of classical physics as described by Isaac Newton).
   B. If the balls are charged objects, this same picture gives an illustration of the conduction of current. All the current in—that is, the balls on the rim of the bowl—is equal to all the current out—that is, the balls falling into the hole. This illustrates charge conservation in our world.
   C. In a second animation, we use the laws of motion in the quantum world. If successive balls are set in exactly the same manner on the rim of the bowl, they don't all travel along the same path as a consequence of the laws of quantum theory. Still, one possible outcome is that all the balls still fall into the hole, which means that, even though the laws of motion are different, charge is still conserved.
   D. Quantum theory allows for a second outcome. Given that the balls can move along different paths, some might never fall out of the bottom. If some of the balls never get to the bottom, charge is not conserved.

E. This *anomaly* was discovered in the 1950s: In a quantum mechanical universe, sometimes charge may not be conserved.

IV. These properties must be incorporated into the view from string theory.

A. If charge is not conserved, then all of the photon has relevance, and this results in a *time-like particle* in the mathematical description of the photon. Unfortunately, time-like particles act like tachyons (although slightly different than tachyons) and have the power to render all predictions of physics nonsensical. Thus, a theory with anomalies makes no predictions about the real world.

B. Any kind of force carrier on the quantum level may have an anomaly, not just the force carrier of the electromagnetic force.

C. The presence of anomalies was worked out by three physicists, Stephen Adler, John Bell, and Roman Jackiw, and the resulting mathematics is called the Adler-Bell-Jackiw theorem.

D. This theorem can apply to mass and energy. As shown by Einstein's great discovery, these may be regarded as different attributes of the same thing. It can also apply to other forces in nature, not just electromagnetism. Both the strong and the weak interaction can be subject to anomalies. And if an anomaly occurs, time-like particles appear for all the force carriers, totally destroying the quantum world and the ability to make calculations.

E. So far, this has been a story about our world from the view of the denizens, but the mathematics of anomalies also occurs in string theory. The vibrations of a string produce, from our point of view, particles. If that's the case, then all the force carriers that we know in our world, which are actually vibrations of a string, can acquire anomalies, and the string itself must be subject to anomalies.

F. Just as the presence of anomalies destroys our ability to make predictions in the real world, so, too, does this apply to the mathematical world of string theory. Time-like particles have exactly the same effect as tachyons: They result in mathematical descriptions in which probabilities can be greater than 1 or less than 0.

V. Around 1982, John Schwarz and Michael Green worked out the mathematics of supersymmetry to banish the tachyon. They also found that they could incorporate, into this concept, forces similar to the electromagnetic and other forces. Because the mathematical world of Schwarz and Green was quantum mechanical, however, it might also be subject to anomalies.

A. While working on the question of the relationship of anomalies to forces and force carriers in superstring theory, Schwarz and Green found something remarkable.

1. Since all forces have force carriers, they all have charges associated with them. The electrical force is associated with electrical charge; the strong nuclear force has color charge; the weak nuclear force has weak charge; and gravity is associated with the Newton constant. If there are many forces at work, then there must be many charges.

2. In looking at the mathematics of something similar to our bowl analogy, Schwarz and Green found that only two mathematical constructs allow all the balls on the rim of the bowl to fall out of the bottom. One of these constructs required exactly 496 charges! From where did this number arise?

3. Recall again the discussion of areas and invariance and our illustration of the Pythagorean Theorem. In two dimensions, the area of the green square is always equal to the area of the blue plus the area of the brown. A similar construct for these areas was given in three dimensions. Each of these areas can be thought of as representing charge, at least in terms of the mathematical calculations of Schwarz and Green.

4. In our world, there are three basic rotations: about the z axis, the x axis, and the y axis. In string theory, one way to arrive at the number 496 is to consider a world in which there are 32 directions to rotate, and then just add dimensions, as in the conception of the Pythagorean Theorem for the original bosonic string. The number 496 is the number of rotations that can be constructed if the world had 32 directions. For this reason, the string that

Schwarz and Green found in 1982 was called the *SO(32) superstring*.

B. The SO(32) string is a way to understand where the number 496 comes from: Almost magically, for 496 charges, all the charges make it to the bottom of the bowl, but for any other number, anomalies appear, which result in time-like particles, which in turn, destroy probabilities. Only using 496 in this special way, considering rotations, does a piece of mathematics emerge that might allow the ability to make predictions about our world. This was the birth of superstring theory.

C. In 1982, it became clear that the goal of Einstein was within reach.

  1. Both Maxwell and Einstein tried to accomplish the same goals using different approaches, and both failed. Maxwell wrote equations to describe electricity and magnetism, then tried to modify his equations to describe gravity. Einstein wrote equations to describe gravity, then tried to modify those equations to include electricity and magnetism.

  2. The magical number 496 arose in the context of a string theory, and the difference within a string theory is that all theories describing closed strings necessarily describe gravity. On the other hand, strings are capable of vibrating in many different ways, and one of their other ways of vibrating describes the force carriers associated with forces like electromagnetism.

  3. A theory that embodies in its mathematical structure the description of photons and the description of gravity implies success where both Maxwell and Einstein failed, and indeed, the world's first unified field theory was born in 1982 with the work of Schwarz and Green and the construction of the SO(32) string.

Readings:

Harmon, *The Natural Philosophy of James Clerk Maxwell*.

Oerter, *The Theory of Almost Everything: The Standard Model, the Unsung Triumph of Modern Physics*.

Questions to Consider:
1. What is charge conservation?
2. What was the problem with charge conservation in the quantum world before 1982, and how did Schwarz and Green solve the problem?

Lecture Thirteen—Transcript
Gauge Theory—A Brief Return to the Real World

In this lecture, we're going to begin by talking about some physics in the real world. String theory, as I emphasized, is not yet science. It's not yet physics; it's mathematics. But the peculiar thing about this mathematics is that it looks like the mathematics that we use to describe the real world. So in some ways, it is mathematics that's inspired by the science that we know, and it actually begins to make predictions that there are things, which ultimately we will want to test, but it's not there yet. So we're going to go back in time, for the beginning of today's lecture, to the year 1861. In that year, a young Scottish physicist by the name of James Clerk Maxwell produced the first color photograph. Now that surprises lots of people because, most often, people think that color photography is a new development, but in fact, this young man had the capabilities and the technology to do this at that point in time. What's really more interesting in this story, however, is that he first figured out how we see color, and that's how he was able to produce the technology.

Maxwell was the physicist upon whose shoulders Einstein most clearly stood in producing the Theory of Special Relativity. Maxwell's 1873 paper entitled *A Treatise on Electricity and Magnetism* contains four equations, and from these equations humanity was led to an understanding of what light is. The equations tell us that light is an electromagnetic wave propagating. No one had ever thought of such possibilities before, and in order for this to happen we needed to have electric and magnetic fields that changed as they moved through both space and time.

They play, in fact, a sort of game of leapfrog. First there's a little bit of an electrical field that has to change in space and time. That generates a magnetic field, which then must change in space and time. Next, an electric field, and this happens over and over again at a certain speed. In fact, the speed is the speed of light. So this was Maxwell's great discovery, and all of our new high technology, all of our communications technology, in fact, can be traced back to these four equations. Sometimes in talking to the public, I call these equations the DNA of modern information technology, because they contain the instructions for how we build things such as radios and televisions and what have you.

Now Maxwell's predictions started off as a piece of mathematics, because remember, I said there were four equations that he wrote. The way mathematics is turned into science is that someone tries to falsify the equations. Science is really not about truths. That's what people often say, is the notion of truth in science. In fact, science is about finding beliefs that are less false, so when someone asserts that an equation applies to our world, what you try to do is prove that it doesn't. This was done in 1888 by a German physicist by the name of Heinrich Hertz, who actually built an experimental device and detected waves of electromagnetic energy that were predicted by the equations; yeah, Hertz as in *Mega-Hertz*. That's where we get the expression from. The Hertz in the word Mega-Hertz is tribute to the work of this physicist.

Now interestingly enough, after Maxwell's great triumph of discovering these equations, he tried to describe gravity. Wow, think of that. You know, this man just figures out how electricity and magnetism work, and then immediately sets the goal for gravity. Unfortunately, in an echo of Einstein's experience, he also failed, but roughly speaking, his attempt occurred about 50 years before Einstein tried to take his theory of gravity and combine it with electromagnetism. So in some sense, the two of them are wonderful bookends in history, in that both tried to achieve the same goal, but from two different points, and both failed.

Now in the description of the photon, I said they have to change in space and time. Let's talk about change a little bit, and for this purpose, I have a video. We've seen this video before. This is the one where we talk about how invariance occurs when you can take a right triangle and look at the respective areas on the side. Well, look at the picture here; it's still, but we can put it in motion, and we can see changes. In fact, we see a change in the area. As the blue area gets smaller, the brown area gets larger, but they do so in such a way that the sum of the two is always the green area. So that's the whole notion of change, but let's talk about change in a slightly different way. This illustration is supposed to be two bowls of ping pong balls, and as you can see, the two bowls are different in terms of how many ping pong balls are in them. One has clearly more ping pong balls than the other. The one on the left is fuller than the one on the right.

Now let's induce a change. The change is that we're going to take ping pong balls out of each bowl, so here we go. But you'll notice that what happens is that, even though the bowls are filled to different capacities, the number of balls that go out was actually the same. So this is actually a very interesting idea because it says that you can take two things that are actually different, but the change in them can be the same. Now the reason this is a very important idea is because this is what Maxwell's Equations tell us about the relationship between electricity and magnetism, which is something that we can measure in a laboratory, and these things that we'll call photons. Remember, we've talked about that quantum of the electromagnetic field. The thing that is the message carrier for the electromagnetic force is the photon.

Well, what Maxwell's Equation tells us is that it is actually the change in the photon that produces the electric and magnetic fields. The equations are even more remarkable because what they also tell us is that if there is no change in the photon, then you don't actually produce electromagnetism. Now our example of the two bowls has a very important lesson, because what they tell us is that it's possible to have two photons where each of them experiences the same change, even though they're different. But because the electric and magnetic fields are produced by the changes, not by the photons themselves, the two photons are indistinguishable. This is a very important idea in physics. In fact, it is so important that it has a special name. We call it *gauge invariance*, the fact that it's the change in photons that produces the things that are physically measurable, not the photons themselves.

Now this idea that the change is what's responsible produces for us a very interesting result, because suppose we imagine that we have photons in a box. Well, you can actually do that. Build a box that has mirrors on the inside and shine a light beam in there; it's full of photons. So what we want to do is ask a very simple question. How much energy is in the box caused by the photons? Well, that's an interesting question because we know that for a given E&M field, which are what causes charges to move or would bend a magnet if we were to put it in the box. We know that for a given E field, more than one photon can produce that. Well, that's interesting because we can imagine two boxes where the electric and magnetic fields are the same, but the photons are different, but their changes are the same. That's the important message.

We can go on to the question of energy. It turns out that the energy that's in each box actually depends on the photons. Now we have a puzzle, because since it's actually the change in the photons that produces the energy, the electric fields, and it's the electric fields that we can measure, then we cannot, in fact, directly measure the photons, and that leads to a very important principle. It's called charge conservation. The fact that you cannot directly tell which photon produces how much energy is the result that leads to charge conservation. This was actually first understood by a mathematician by the name of Emmy Noether. In fact, there's a very great result in mathematical physics that's called the Noether Theorem. What it tells us is that if you have some quantity that changes, but that physically measurable quantities don't depend on those changes, then there must be conserved currents.

Well, let's stop and think about this for the photon. We're talking about these changes, and we know that we can't quite measure photons directly; we can only measure the change in them. So what's the conserved quantity associated with that? Well, remarkably enough, it's the electric charge. You've probably heard in your science class that charge is a conserved quantity. Well, it is. But it turns out that because of Noether's Theorem, we understand that the conservation of electric charge is actually a lesson about the ability of two different photons to produce the same electric or magnetic field. So we've learned a very great lesson here; that it's not, in physics, always the quantity that's of interest. Sometimes it's the change in the quantity that's of physical interest.

So now let's talk about the fact that our world is quantum mechanical. Remember, we have, in previous lectures, actually gone down in size, leaving our world and going down to the world of what I call the denizens of the quantum world—the quarks, the leptons, the photons, the intermediate vector bosons, the force carriers. Those were the denizens we met. In their world, the laws of motion are very different from the laws of motion in our world. Now that has an important implication for charge, and in order to show this implication, I have another image that will be useful. Here's a bowl, and as you can see, there's a ball that's on the side of the bowl. Now what we're going to do is to watch what happens as we put this in motion. We're going to let the ball roll down into the interior of the bowl, and you'll notice there's a hole down there. So it rolls down and falls through the hole.

If we put a second ball in the same place, it does exactly the same thing along exactly the same path. We put a third ball there. It does the same thing. Now the fact that all of these balls do the same thing if you put them in one place is a result of classical physics. Classical physics says that if you put an object in a certain position in the same way, it will always follow the same path, and that's what we see in this idea. Now imagine these balls are actually charged objects. Then what this is a picture of is the conduction of current, and we can see that all the amount of current we put in—that is, putting the balls to the side of the bowl—goes out the bottom. So all the current in is equal to all the current out. That's actually charge conservation. That's what we physicists mean by charge conservation.

Okay. So that's how charge conservation works at the classical level, at the level in which we're used to looking at our world. But remember, in the quantum world, things are different. So let's go visit the quantum world and ask about the same experiment. Here again, we see our bowl with—this time we're going to put several balls on this lip of the bowl, and there are our balls. Let's put them back at the same place so they start at the same place. Now let's watch what happens as they roll down into the middle. Well, as you can see, they don't all travel along the same path. In fact, they can travel along different paths. Now why is that? Well, let's remember how the quantum world actually works.

In the quantum world, we cannot say with definite precision what exactly happens between the time we observe something at one time and observe it at a later time. So let's go through this again and see if we can figure out what's going on with our balls. So we would observe the ball when we put it on the lip, and then as the balls move along, you'll notice that some of them don't follow the nice classical path. Some of them, in fact, follow other paths along their way to the hole. So we can put one ball on the side of the bowl, but as it moves down towards the hole, it can take many different paths. In my illustration here, you can see this occurs by having the balls follow different paths. I have a green ball, a white ball and a red ball, and they—some of them wander around as they try to make their way toward the hole at the bottom of the bowl. This wandering—that's quantum mechanics at work, because even though we put the balls in the same place at the same time, according to the laws of quantum mechanics, they're allowed to wander around in trying to reach the bottom.

So now we can see the wandering. Some of them go to one side, some to the other. Eventually, all of them go out the bottom of the hole. Now this contains an important lesson. Even though the laws of motion are different in the quantum world, as long as all the balls that are placed on the lip of the bowl get out the bottom, charge is still conserved, even though the world is quantum mechanical with all the wandering in the middle. Well, is that the only outcome that's possible? Well, the answer, interestingly enough, turns out to be no. There's one other outcome, which is actually very startling. Since the balls can wander, what would happen if we put a number of balls on the same place, one after the other, and watched them wander? But suppose some of them never wandered out of the hole in the bottom. We put them in the top, but they didn't get out the bottom. What would that mean?

Well, since it's actually the—for every ball that you put in the top must go out the bottom—that's what we mean by charge conservation. If some of the balls never get to the bottom, it means charge is not conserved. This is such a startling result that even physicists had a hard time accepting this. This was actually first discovered in the 1950s, that in a quantum mechanical universe, if you watch how charge propagates, sometimes it is not conserved. So here we see an example where, if we watch our balls, instead of all of them going down to the middle and going out of the hole, some of them just wander around the hole forever and never go out the bottom of the hole. That's what physicists call an anomaly. Now all of the things I've just told you are properties of our world, and we have mathematical descriptions to describe these things. So now we want to take some of these properties and put them into our picture of string theory.

So how should we do this? Well, we've learned about anomalies. We've learned about the fact that sometimes things that are physically measurable are actually related to changes as opposed to being directly related to the objects present. When anomalies were first discovered, as I said a moment ago, even physicists had a hard time accepting that. In fact, there's a story. Let me tell you a story about this. Supposedly, there was a graduate student who made a calculation and found that when he did the calculation one way, he saw charge conservation. He did the calculation a slightly different way; he concluded that charge was not conserved.

At first, because of this, he was convinced that he had made an error in the calculations. So he checked the first method of calculation and found out again that charge was conserved, checked the second method of calculation and found out that charge was not conserved. And this young student, who had hoped to become a theoretical physicist, concluded that whatever was going on with his calculations, he wasn't wrong, and he became an experimental physicist instead of a theorist. Now I don't know if the story is true, but one often hears this story about the discovery of anomalies and what it has done.

So anomalies are truly bad news. Why? Well, you see, if charge is not conserved, then it means that the part of the photon that doesn't change has relevance. But if that's true, then there's a part of the photon, at least for the mathematical description of the photon, which is an object we physicists call a time-like particle. Now that name may not scare you, but the only problem is time-like particles act like tachyon. In our discussion, we have seen that tachyon has the power to destroy physics. It renders all of our predictions nonsensical. Time-like particles have exactly the same property, even though technically they're slightly different from tachyon. So if you have a theory with anomalies, it turns out that is a theory that makes no predictions about the real world. Now the interesting thing about anomalies and tachyons is that the anomaly doesn't just have to apply to the force carrier for the electromagnetic field. Any kind of force carrier potentially—when you look at it at the world of the quantum, the world where the denizens run around—any kind of force carrier might have an anomaly.

The presence or absence of anomalies was worked out by three physicists, Adler, Bell and Jakiw. They gave the mathematical criteria, which basically said that for every ball that you put at the top of the bowl, it goes out the bottom. That's when charge is conserved. But they also found out what conditions had to be satisfied when you put a ball at the top, and repeatedly do that, and find some of them that wander around the bowl and never go out the bottom. It's called the Adler-Bell-Jakiw Theorem. It can apply to mass and energy because Einstein's great discovery was that energy and mass are the same thing. It can apply to the other forces in nature, not just electromagnetism. We've talked about the two forms of nuclear energy, the strong interaction and the weak interaction. Each of these can be subjected to anomalies, and if an anomaly occurs, time-like particles appear for all the force carriers, totally destroying the quantum world and our ability to make calculations.

Well, the thing that's really interesting—because this, so far, has been a story about our world—is that the mathematics of anomalies actually occurs in string theory. You remember, string theory is really a mathematical construct. It tells us that at very small levels, at the very tiniest levels, our universe looks as if it's somehow constructed by little pieces of pasta, little filaments. The mathematics of that is what we string theorists actually study. We study how these things are described in terms of their mathematics, and it turns out that, as we've discussed, when a string vibrates, that vibration is too small to actually produce sound. So instead the vibration, from our level, appears to be a particle. So if that's the case, then it means that all the force carriers that we know in our world—which are actually, from this point of view, vibrations of the string—it means that if each of these force carriers, or any of them, acquire anomalies, then the string itself must be subjected to anomalies.

So if the string is subjected to anomalies, what does that actually do? Well, just as anomalies destroy our ability in the real world to make predictions, the presence of an anomaly in the mathematical world of the string has precisely the same effect. We find that objects that are very similar to tachyon appear. We know that the effect of tachyon—or the tachyon, more properly—is that it destroys the notion of probabilities. Time-like particles do exactly the same thing. You have a mathematical description where probabilities can be greater than one or less than zero, and of course, we have no way to interpret such a thing. So let's now pick up the story about strings because around 1982, a marvelous discovery was made. We talked about the fact that strings come in several different varieties. When you add spin to strings and put a certain restriction that says that you want equal numbers of fermions and bosons to appear in the oscillatory patterns, you get a property called super symmetry.

Well, in 1982, two physicists, John Schwarz and Michael Green, actually worked this out for the first time, how to see super symmetry and use it to permanently banish tachyon. They actually did more than that, because they also started to ask the question can we incorporate forces that are like the electromagnetic force, that are similar to the weak and strong nuclear forces. They found out in their mathematical world that yes, they could actually do that also. But this mathematical world is a world that's quantum, which is in the quantum regime, and therefore might be subject to anomalies. In 1982, while in Aspen—there's a physics center in the city of Aspen where typically, for the last 20 years, you can find some of the country's and the world's most active physicists working feverishly to derive new results. Well, in 1982, Michael and John were there, and they studied the question about whether this idea of anomalies had anything to do with the forces and the force carriers that they had constructed for the so-called superstring. What they found was quite remarkable.

All of the forces, as we've talked about, have force carriers, and all the forces also have charges associated with them. With the electrical force, it's the electrical charge. For the strong nuclear force, we call that the color charge. For the weak nuclear force, we call that the weak charge. So every single force is always associated with a charge. You might wonder what the charge for gravity is; it's actually Newton's Constant. So every charge is associated with force. Now if you have many, many forces, it means you have to have lots and lots of charge. So in 1982, Green and Schwarz did a calculation very similar to our bowl analogy. They looked at the mathematics of the theory. They started putting little balls in, and then they figured out which theories have the property that all the balls that you put in the top of the bowl get to the bottom.

What they found out was that only two such mathematical constructs exist. In one of these constructs, they found that you need exactly 496 charges. Where in the world would you get such a number from, 496 charges? Well, since string theory is about mathematics, and since by now those of you who have been listening to me have actually, in some sense, peeked into PhD-level training for mathematics, we can explain this number, 496. Remember, when we go back to talk about our areas, and we'll do that one more time, bring that back. This is a very useful concept. When we talked about our areas and the invariants, we could do it in two dimensions where the statement was that the green area is always equal to the sum of the blue plus the brown. That's true no matter how we orient the two areas, so we can make the brown area larger or we can make it smaller, but the sum of the two is always the same. Or we could do something else. We could actually take the areas and construct a third one, which was, as we can see here, in terms of my ladder, it means I've tilted the ladder over.

Now each of these areas can be thought of as like charge, at least in the context of the calculation that Green and Schwarz did. So how do you get the number 496? Well, the answer turns out mathematically to be the following. You see, in our world, we can do rotations, so I can rotate like this. That's what we physicists like to call a rotation about the Z axis, because Z is up for us physicists. Or if I hold my hand like this and did a backflip, I would rotate about my arm. That would be a rotation about the X axis. Or I could hold this arm out at 90 degrees with respect to the first arm, and once again do a back flip. That's what we would call a rotation about the Y axis. So in our world, there are three basic rotations. In the world of the string, one way to arrive at the number 496 is to consider a world in which there are exactly 32 ways in which to rotate, because this number, 496, is the number of rotations that you could construct if the world had 32 directions.

So for this reason, the string that Green and Schwarz found in 1982 was called the SO(32) string, because it's a way to understand where this number, 496, comes from. Why is it, magically, that for 496 charges, all of them make it to the bottom of the bowl; whereas, if you pick any other number, you always have an anomaly? If you have an anomaly, you have time-like particles. If you have time-like particles, that destroys probabilities. So only through this magical number of 496 do you have a piece of mathematics that you can use to make predictions about our world. This was the real birth of superstring theory, and in 1982, it became clear that the goal of Einstein was within our grasp.

Now what was this great goal of Einstein? Well, early in this lecture, I talked about the fact that both Maxwell and Einstein actually tried to do the same thing, but from two different points. Maxwell wrote the equations that described electricity and magnetism, and after they were successfully written, he wanted to include gravity. Einstein, on the other hand, as we've learned in a previous lecture, wrote the equations that described gravity. After he was successful, he wanted to modify those equations to include electricity and magnetism. So both of them wanted to do the same thing, and both of them failed. In

Einstein's case, after a 30-year-long search, the last 30 years of his life, he still did not have this single description of both gravity and electricity.

Well, the work of 1982 made it clear that we could grasp this thing that we were almost at the goal of Einstein, because this magical 496 that I was talking about was in the context of a string theory. Now the difference about a string theory is, if it's one of the closed strings, one of the ones that look like a little hoop that we've seen in previous animations, all such theories necessarily describe gravity. On the other hand, the string is capable of vibrating in many different ways, and one of its other ways of vibrating describes the force carriers that are associated with things such as electromagnetism. So if you could find a theory that had in its mathematical structure the description of photons as well as the description of gravity, then you have succeeded where both Maxwell and Einstein failed. So it was clear that it was within grasp, and when the SO(32) string was completed, that grasp became not a potentiality, but an actuality.

The world's first unified field theory was born in 1982 with the work of Green and Schwarz and the construction of the SO(32) string. Now this was a source of great joy in the physics community because it meant that this 30-year-long quest of Einstein had been fulfilled. By the end of Einstein's life, most of the physics community thought of him as kind of passé, someone who was off on some journey that had no useful purpose to physics. Very few people, by the end of Einstein's life, followed his quest for the unified field. Einstein died in 1955. It was 1984, some 29 years later, when his dream was realized in Aspen, Colorado.

Lecture Fourteen
Princeton String Quartet Concerti—Part I

Scope:

Humans are generally familiar with the properties of light. It has brightness (*intensity*) and color (*frequency*, or *wavelength*). Maxwell's great discovery that light is a kind of wave implied one additional property. In this lecture, this property of light, *polarization*, is discussed, and we will see how an analog of this property was used to construct a new kind of string.

Outline

I. A computer graphic shows light as a wave combining two different oscillating "objects"—more formally, *fields* (electrical and magnetic fields).

   A. In the animation, the blue balls on the horizontal axis constitute an electrical field; the green balls on the vertical axis constitute a magnetic field. As a wave of light propagates, the two fields oscillate together. Thus, light is a dual vibration of electricity and magnetism.

   B. Light has another property called *polarization*. To simplify this explanation, our discussion will ignore the magnetic field and concentrate only on the electrical field.

      1. We see an animation of a light ray with the characteristic shape of a wave. A crest appears on the left side of the image and a second crest on the right side. The distance between these two crests defines the *wavelength*, known as its color.

      2. The tip of an electrical field of a light ray can be described as it enters the eye of an observer. The electrical field lines might simply vibrate up and down. This is called *linear polarization*. The field lines might also rotate, which is called *circular polarization*. This can occur in either a counter-clockwise or clockwise sense.

      3. In the center of an animation we find a large circle with a line oscillating vertically to represent linear polarization. In linearly polarized light, the electrical field vibrates only along this single line. In the two

smaller circles, the green arrows rotate around the smaller circles to illustrate circular polarization.
   4. The property of polarization was first discovered in the mathematical description of light, but most people cannot detect polarization with their eyes. In 1846, a geologist named von Haidinger noticed he could correlate a particular image with the polarization of light. This effect is called *Haidinger's brush*.
   5. When people who are sensitive to polarization look at a bright sky, they see a diffuse pattern that looks roughly like two figure 8s—one vertically oriented and one horizontally oriented—caused by the polarization of light. Turning one's head to the side may cause the figure 8 to rotate, as shown in the animation. If the light is linearly polarized, the figure 8 doesn't move.
   C. There are many devices to detect polarization. The polarization properties of light are used in polarized glasses to reduce glare.
   1. An animation depicts an electrical field as balls lined up along an axis. If the electrical field is set in motion, a wave starts to form, but an object is blocking the wave. In this orientation, the light cannot pass the blocking object. If the orientation of the object were changed to exactly line up with the electrical field, the light could pass. This is the basic mechanism behind polarized glasses.
   2. The reason polarized glasses cut down on glare is that glare typically comes off a surface with horizontal polarization; in other words, the electrical field is moving horizontally. The lens, then, can be oriented to block the horizontal polarization.
II. The rest of this lecture will show how an analog of Haidinger's brush was used, in 1984, to construct a new string.
   A. Green and Schwarz looked at the mathematical properties associated with anomalies in the superstring and found the anomalies could be eliminated with exactly 496 charges.
   B. This 1984 construction, which allowed for anomaly freedom in string theory, also presented a puzzle.

1. As stated in the last lecture, 496 is the number of rotations that one can perform in a mathematical world in which there are 32 different directions. Green and Schwarz also realized that, in addition to the rotations around the 32 axes, there was another way to arrive at the number 496.
2. This other way was discovered by Élie Cartan, a mathematician who studied rotations and certain mathematical generalizations. In trying to find a description of the most general mathematical rotation, Cartan also found the most complicated rotation that one can describe using mathematical formulations. This rotation is called E_8; the *E* stands for *exceptional*.
3. Rotations are described by angles; therefore, if we have 496 rotations, we must have 496 angles.
4. The mathematical object E_8 is associated with exactly 248 rotations and, therefore, 248 associated angles. It is easy to see that 248 + 248 = 496, the magical number that prevented the anomalies from causing havoc in superstring theory. One set of rotations had been found, but where was the other set?

III. Let's return for a moment to the SO(32) string, which played an important role in understanding string theory.
   A. Where exactly on the strings do these 496 charges occur? For the SO(32) string, the charges occur at the ends of the open type of string. Before 1984–1985, this was the only way that physicists knew how to describe a string that had charge; namely, the string carried charge at the ends of the filament.
   B. Why must the charges be restricted to the ends of the filaments?
      1. Some lectures ago, we noted the fact that a closed string has properties similar to a sound wave in a pipe. Recall that animation showed a sound wave as a bowed shape tied down at the ends of a closed pipe. In an open pipe, the ends of the wave were free to move. In string theory, this same mathematics occurs.
      2. The closed pipe corresponds to a string with the ends tied together. These kinds of strings always describe

gravity. The open strings, the ones that have the charges on the ends, always describe photons, but they never describe gravity.
3. A theory that describes both gravity and charges seems to require a string that combines both open and closed types, knowing that charges occur only on the open strings.

C. In 1984, four physicists at Princeton University, Gross, Rohm, Martinec, and Harvey, who became known as the Princeton String Quartet, solved this puzzle. They became mathematically conscious that there was a kind of brush associated with open and closed strings. To understand their work requires a second look at open and closed strings.
1. An animation shows a closed string in motion. It oscillates, but there are points, called *nodes*, at which the string is not moving.
2. The Princeton String Quartet realized that every mathematical description of closed strings written up to 1984 had the property that the nodes of closed strings remained fixed at the same locations.
3. These fixed nodes are the analogs of linear polarization for real light rays in our world (the middle circle in the animation of polarization), where the electrical field vibrated vertically but never rotated.
4. Understanding these nodes and their connection to linearly polarized light prompts a question: Is it possible to introduce into the mathematics of string theory objects that correspond to circularly polarized light? If so, it turns out that charges on the string need not be restricted to its ends but could be distributed throughout the length of the string.
5. Consider a second animation of a string, with the nodes indicated by red dots. In this animation, the nodes move. The motion of the string roughly corresponds to left circularly polarized light; thus, the string is called a *left-mover*. Of course, we can also write the mathematics to describe *right-movers*.

D. The Princeton String Quartet went a step further by writing the mathematics that took advantage of the second set of Cartan's rotations and found that the number 496 comes from two E_8s. This *heterotic string* can be interpreted as the logical combination of strings from different dimensions. The left-moving parts of the string describe a superstring (meaning it has spin associated with it), but the right-moving parts of the string describe the old bosonic string that had the tachyon problem. Remarkably enough, this synthesis works.

   1. The supersymmetric part of the string, the left part, eliminates the tachyon, but the right part of the string is like an open string, which allows the charge to be distributed throughout the heterotic string.
   2. Many physicists questioned how this synthesis could be accomplished mathematically, but in fact, the mathematics already existed in the physics literature, provided by two physicists, Makoto Sakamoto and Warren Siegel. The first one gave the mathematical description of *heterotic fermions*, and the second gave a different mathematical description—that seemed unrelated—of *chiral bosons*.

IV. The description the Princeton String Quartet wrote to measure its energy does not contain any quantities that add to 496. Where were the charges?

   A. This part of the story is linked to a Scotsman, John Scott Russell, who in 1834, observed a strange solitary wave in a canal. This kind of wave, called a *soliton*, has some properties of particles; namely, two of these waves can bounce off each other.

   B. The Princeton String Quartet argued that the charges that make up the magical number 496 are distributed between the ordinary waves of their theories and the mathematics that describes solitons.

Readings:

Darrigol, "The Spirited Horse, the Engineer, and the Mathematician: Water Waves in Nineteenth-Century Hydrodynamics," in *Archive for History of Exact Sciences*, 58.

Greene, *The Elegant Universe* and *The Fabric of the Cosmos: Space, Time and the Texture of Reality*.

Kaku, *Hyperspace: A Scientific Odyssey through Parallel Universes, Time Warps, and the 10th Dimension*.

Questions to Consider:
1. Describe linear and circular polarization of light.
2. Explain the analog of polarization in string theory.

Lecture Fourteen—Transcript
Princeton String Quartet Concerti—Part I

In a sense, humans are creatures of light. I mean in order to see this, many examples suffice. One, for example, is to go to the beach on a sunny day. You'll find it full of people. Lots of us are familiar with the properties of light. It has brightness and color. Scientists call the former intensity and call the latter either frequency or wavelength. Maxwell's great discovery was that light is a kind of a wave. We have a few graphics to show us some of this. It's a kind of wave, but the thing that's interesting is that it combines two different oscillating objects, objects in quotes there, or more formally, fields. There is an electric field and a magnetic field associated with light. Remember, I told you it was a game of leapfrog, so you have to talk about both of them.

On this animation, what I've done is tried to show you these two objects, these two entities, these two quantities that we call fields. The blue balls, which you can see laid out along the X axis, you can think of as electrical fields. The green balls, as you see laid out along the Z axis, which is up and down you can think of as a magnetic field. Then as a wave of light goes by, the two of them actually oscillate together. There you can see the electric field vibrating in this plane and the magnetic field vibrating like that. So that's what light is. It's a dual vibration of electricity and magnetism. That was Maxwell's great discovery.

Well, it turns out light has another property called polarization. Polarization is a little bit more difficult to wrap our minds around, as some people say, so we're going to do this in several steps. Again, in order to simplify things, I'll forget about the magnetic field and only talk about the electrical field. So here's another picture of a light ray. If you look at this, you see it has the characteristic shape of wave. You'll notice that there's a crest on the left-hand side of the image, as well as a crest on the right-hand side. If you go the distance between these two, that's what scientists call the wavelength. Interestingly enough, for light, that is exactly the definition of color. If you know the crest-to-crest distance of a light ray, you know its color. For humans, we can only perceive light when it's in a particular range. So outside of that, the light is there, but we can't see it. But within this certain range, we can, and that's typically the colors of the spectrum in which we see.

Now this property called polarization, we're going to approach in two different ways. Let's imagine watching the electrical field as it comes toward us. One thing that it can do is simply vibrate up and down, and then that's what we scientists call linear polarization. Another thing that it can do is actually more interesting, which is what we call circular polarization. When a circularly polarized light ray approaches us, if we watch the electrical field lines, they rotate around. In fact, I have an animation to show both of these, and here they are. In the middle, we're going to watch linear polarization. What we'll find is that the blue line will simply go up and down, and that's what linear polarized light is; the electrical field vibrates only along a single line. The two green arrows that you see in this illustration, however, will wind around as we watch them, and that's what we mean by circular polarization. The electrical field line, in that case, winds around. So here we set the movie into play, and there you can see the linear polarization with the blue line going up and down. But each of the green lines moves around in a circular path.

Now this last property of polarization is there in the mathematical description of light, but most of us cannot detect polarization with our eyes. But interestingly enough it turns out, that some people can actually see polarization. This was first discovered by a geologist by the name of von Haidinger. In 1846, he noticed that he could actually correlate a particular image with the polarization of light. This effect is now called Haidinger's Brush, and it's due to some peculiar nature of the structures of the human eye. In fact, cameras cannot see polarization, but people can, and it's even a skill that you can learn. So it turns out that the way that Haidinger's Brush works for those who can see it—and let me tell you, I'm not one of them, but I've read the phenomena—is that when someone who is sensitive to polarization actually looks at a bright sky, they will see a diffuse pattern that, roughly speaking, looks like two figure eights. One of them is vertically oriented; the other of them is horizontally oriented, and that figure eight shape is actually caused by the polarization of the light.

When such a person looks at a bright sky, if they turn their head to one side, it will seem as though that figure eight will rotate one way. That will be caused, for example, when the light is polarized, such as we see in the circle that I'm pointing to in the image. Let me set it in motion to remind you what happens. So there we see one of the polarized beams. So for a person sensitive to polarization, when they turn their head to the side, they actually see this figure eight rotate one way. If the light is polarized in the other sense, and they turn their head to the same side, the figure eight tilts to the opposite direction. That would correspond to the circular polarization that I'm indicating with my pointer here, and if the light is linearly polarized, then the figure eight doesn't move at all. They turn their head, and it still keeps the same orientation, and that corresponds to the center image where we see the electrical field associated with the light beam simply vibrating up and down.

This is quite remarkable. In fact, before I prepared this course, I was not aware of this phenomenon—that there are actually people who are sensitive to the polarization of light, and I think most of us are totally unaware of this optical phenomenon. But apparently, some people are. So that's what polarization of light is. It's actually detectable by a small number of us. We have many, many devices for detecting polarization. In fact, polarization actually is something that we can take advantage of in constructing things such as polarized glasses. Let me show you how that works. Here we have, again, a depiction of our electrical fields. It's the balls that you see lined up along the axis that the pointer is indicating. And now if we set an electrical field in motion, you'll notice a wave starts to form. But we have an object that's blocking the wave, and when that happens, the wave never makes it through the object that's doing the blocking. In terms of light, that means that if you think of this light coming in, if we have this orientation that blocks it, then the light that goes through doesn't make it.

On the other hand, we can also say well, let's rotate the object doing the blocking; so here we see this. Again, we set up a light ray. It comes through, but because the orientation of the blocking object is exactly lined up with the electrical field, the electrical field or the light makes it to the other side. Now although these are simple models, we actually know how to build these things. They're called Polaroid glasses. The reason Polaroid glasses cut down on the amount of glare that one sees in the day is because glare typically comes off of a surface with what's called a horizontal polarization. That means that the electrical field line is moving like this. So if you set up your opening in your blocking so that it goes like this, it means that light doesn't get through, and you don't see the glare. So this is how Polaroid glasses actually work.

Now, of course, all this is real physics, and several times during this presentation, I'm sure you have found yourself asking why in the world is this guy telling us stuff that has nothing to do with string theory? Well, the reason is because every single physical phenomenon that I have described, in fact, has an echo, has a remnant, which shows up in string theory. Today's lecture is going to be about how the analog of Haidinger's Brush was used in 1984 to construct a new string. You will remember, at the end of our last lecture, we talked about anomalies. Anomalies were phenomena that occur because the world of the quantum is complicated, and motion can occur in ways that we could never see in the world around us.

For currents or charges that move in the quantum world, if you put some charge into a wire, you would like to get the same charge out. That's called current conservation or charge conservation. But in the quantum world, because the objects are free to wander, not all the charges you put in come out, and when that happens, we have an anomaly. An anomaly, as we discussed in the last lecture, was a horrible thing to have around. So Green and Schwarz looked at the mathematical properties that were associated with anomalies in the string, the superstring, first constructed explicitly in 1982, and they found an amazing result—that if you have exactly 496 charges, then the string would have no anomalies. It would be as if we put 496 different colors of balls in our example, and every time you put one of the balls in, for every ball you put in, one would come out. That's the current conservation. That's the great discovery of the superstring.

This 1984 construction, which completed the anomaly freedom of the string theory, actually led to a puzzle. In my last discussion, I explained to you, in some sense, where the number 496 appears in mathematics. It's the number of rotations that one can perform in a mathematical world in which there are 32 different directions about which one can perform rotations. Now if one looks at that, at an earlier stage, well, we were talking about rotations. In some lectures ago, I talked about the fact that there's a particular form of mathematics that's exactly associated with the idea of rotations, or more generally, what physicists and mathematicians like to call groups. So when you have something like 496 that works for a particular set of rotations, you can go often to the mathematical literature to study the properties of these things. Toughly speaking, this occurred after Green-Schwarz anomaly freedom was discovered.

In fact, they realized almost as soon as they wrote the paper that in addition—in fact, the comment is actually in the same paper—that in addition to the rotations around these 32 axes, which gives you this number, 496, there's another way to get the number. Now this other way was actually discovered by a mathematician. In fact, it's the mathematician Cartan, who we've mentioned several times previously in these discussions, who was really the person who pulled together the idea of how to describe mathematical objects that are the most general possible, that are similar to rotations. When Cartan actually solved the problem of finding the most general mathematical rotation, he observed that there were lots of different possibilities, but there's one possibility that's extremely exceptional. In some sense, it's the most complicated rotation that one can describe using these mathematical formulations.

This particular set of rotations goes by a name; it's called E8. The E stands for the word exceptional, and the interesting thing about E8 is that you can—one thing about rotations to keep in mind is that rotations are described by angles. When you rotate something, you can always talk about by how much angle I rotated. So if you have 496 rotations, there must be 496 angles that tell you by how much the rotation was performed. Well, back to the story of E8. It turns out that the mathematical object E8 has exactly 248 rotations, and therefore angles, associated with it. Well, you don't have to be Einstein to figure out that 248 plus 248 is 496, but 496 was the magical number that prevented the anomalies from causing havoc in string theory, in superstring theory. It was, therefore, a puzzle in 1984 that if we found this one set of mathematical rotations for which the superstring was a consistent theory, then why not the other set? This led to a very great puzzle.

Now let me say a little bit about the SO(32) string, because it has played an important role in understanding string theory. We described how the string itself is like a filament. But one thing about the SO(32) string is that if you ask about these charges, which I described earlier, these 496 charges, one physical question you can pose is well, where exactly on the strings do these charges occur. The answer turns out to be rather interesting, because for the SO(32) string, what one has to do is to consider both open and closed varies, and you can put the charges at the end of the open type of string. Now before 1984 or 1985, this was the only way in which physicists knew how to describe a string that had charge; namely, it carried charge at the end of the filament. The middle of the filament never carried charge. So there's something interesting going on here, because now we get into the business of whether the string is a little hoop or whether it's a filament, because the charges are always at the end of filaments.

Now that's kind of strange, because why must the charges only be restricted to the ends of filaments? In fact, one of the really interesting parts of this question is that a moment ago, we talked about light—how you could have linear polarization where the electrical field simply goes up and down, or you could have circular polarization where the electrical field seems to rotate in either a clockwise or an anti-clockwise sense. This is actually related to string theory. Some lectures ago, I talked about the fact that a closed string actually must have the properties that are very similar to having a sound wave in a pipe. Perhaps you remember that animation. If you have a closed pipe, and you look at the average displacement of the air molecules at the ends of the pipe, they're not allowed to move, so you see a bow shape that is tied down at the ends. If you have an open pipe, the ends are free to oscillate back and forth.

In string theory, this exact same mathematics occurs. The closed pipe corresponds to a string whose ends are tied together. These are the kinds of strings that always describe gravity among the particles that seem to occur as specific oscillations. The open strings, the ones that have the charges at the end of them, always describe things like photons, but they never describe gravity. So we have a picture here, then, that if you want to describe a theory with both gravity and with charges, it seems that you have to have a string that combines both open and closed. Since the open strings have the charge at the end, that's the only place where you find the charges.

In 1984, a remarkable occurrence happened at Princeton University. That discovery was led by David Gross, a recent Nobel laureate in physics, but for a different work, not this work, which we will describe now. It is as if David Gross, Emil Martinec, Ryan Rohm and Jeff Harvey were mathematically able to perceive Haidinger's

Brush. Remember I talked about polarization. Not many of us can actually see it, but there are some people who can. Well, in some sense, what happened in 1984 was that a group at Princeton, these four physicists at Princeton, became mathematically conscious of the fact that there was a kind of brush associated with open and closed strings, particularly with closed strings, and they used this to their advantage. By the way, this group of physicists, since there are four of them and they were at Princeton, we now refer to them often as the Princeton String Quartet.

So what did the Princeton String Quartet actually do? Well, in order to understand their work, we need to go to look at closed and open strings. So here's a picture of a string. This particular picture I have is of a closed string. In a moment, we'll set it into motion so that we can actually watch it moving. So this is a closed string. It's some kind of a loop, and we imagine that we've deformed it. Now let's set it into motion. There it goes. As you can see, it's oscillating back and forth. But one of the interesting things about these oscillations is that, as you can see where my cursor is right there, the string is actually not moving at that point. That point of no motion is what we physicists call a node of vibration, a point that's fixed even though the extended object is moving. As you can see in this picture, there's not just one node. So here's one at the cursor here. There are other nodes, so let me go to another such node. You'll notice I can put my cursor there, and I'm always touching the string even though it's vibrating. So this particular string, this closed string, the one that is mathematically described by this picture, has nodes in it. Now there's something very interesting about those nodes. They all sit at the same place as we watch the string vibrate.

What David Gross and his collaborators realized was when you looked at the mathematics of closed strings that had been written up to 1984, every mathematical description of a string had this property that the nodes of the closed strings always sat at the same location. Well, once you have the idea that these nodes sit at the same location, roughly speaking, that is like looking at light—so here's our light once again—like looking at polarized light and realizing that this is the analog of the middle picture here, where the electrical field simply goes up and down, but it never actually rotates. It only touches the sphere at two places; either at the top or, when it rotates all the way down to the bottom, it touches at the bottom. But those are the only two places. So in the first figure, if you ask where does the light touch the circle, it's at two places. One's at the top. One's at the bottom. That's like the nodes that we saw in string theory.

So once you know these nodes are here and understand the connection to polarized light, the next most natural thing to do is ask, in the mathematics of string theory, if it is possible to introduce objects that correspond to polarized light. In other words, whether we can find mathematical objects where, when we watch the electrical field lines, what we see is that they rotate around, as the upper two examples do, as opposed to just touching the circle at a single point. Now this is actually important for string theory because, you see, if we can find such a construct, then the charges on the string do not have to be just at their ends. The charges can be distributed throughout the length of the string. So David Gross and his collaborators set out to find such a construct. Of course, since we are doing the mathematics by pictures as opposed to actually writing the equations, it's pretty easy for us to figure out what we need.

Here's such an example. Here again is a string very similar to our first. Notice I've indicated the nodes with red dots in this diagram. Now let's set this one into motion, and you'll notice the nodes move in this one. That's different from the closed string. The nodes move, and in fact, we call such motion left movers or left-handed motions. So that, roughly speaking, corresponds to a kind of left circularly polarized light. In a similar manner, if you can write this mathematics, then gee, maybe you should be able to write the mathematics which describes right movers. So in this famous paper of 1984, indeed the Princeton String Quartet was able to describe right movers, where the nodes move in the opposite sense, so a new kind of string was born.

Now you'll remember I talked about the mysterious number 496 and how, in the original construct of Green and Schwarz, they saw that SO(32) —that is, rotations in 32 dimensions—would lead to this number. Well, what Gross and company did was to write down the mathematics that took advantage of Cartan's second observation about how to get the number 496. You can get it from two E8's; 248 plus 248 equals 496. This construction has a very interesting interpretation because you can also interpret this kind of construction, which is called the heterotic string, as being the logical bringing together of strings that live in different dimensions. That is, you imagine that the left-moving string is a superstring. We've talked about what that means; it has spin associated with it. But all the right-moving parts of the string are associated with the old bosonic string that had the tachyon problem.

Remarkably enough, this synthesis works. And when you study this synthesis, you find out that the problem with tachyon is banned because the super symmetric part of the string, which is sort of the left part, actually gets rid of tachyon. On the other hand, the right-hand part of the string is like an open string, and that allows the charge to be distributed throughout the entire heterotic string. So heterotic string theory was this amazing synthesis that truly set off a revolution after it was born in the period 1984 and 1985. If you look at the physics literature, you'll find an enormous number of citations in the literature about this.

This bringing together—gee, that sounds like a strange sort of thing. It certainly was very startling because when it was done, no one first thought it could be done. In fact, I remember being on a visit to Berkeley talking with a friend of mine, who is a mathematical physicist, who said well, gee, how can you actually do this, this business of making the SO(32) and the E8 work? Fortunately, this bringing together mathematically had the parts necessary lying in the physics literature. Two physicists, one by the name of Sakamoto and another by the name of Siegel, had provided precisely the mathematics that tells you how to bring these parts together. One of these, the one description by Siegel, uses a piece of math we call *chiral bosons*. That basically means a mathematical object that describes something that spins around, and when you put these things together, you find that you can describe the heterotic string.

Well, not quite. There's actually a little bit of a puzzle here. Remember I kept talking about 496, how that was the magic number of charges? Well, in fact, if you do the description, if you look at the description that Gross and company wrote in 1984, you'll find out that they don't actually introduce 496 of anything. So then you have to wonder well, where are the charges that we're talking about? This part of the story has a very interesting connection to a gentleman, a Scotsman, in fact, who's out riding his horse in 1834. His name was John Russell. One day while riding at the Union Canal near Edinburgh, he observed a single wave that formed. In fact, let me read his description of this.

He said, "I was observing the motion of a boat, which was rapidly drawn along a narrow channel by a pair of horses, when the boat suddenly stopped—but not so the mass of water in the channel which had been put in motion by the boat. It accumulated around the prow of the boat and then rolled along with a great velocity, assuming the form of a large solitary elevation; a rounded, smooth and well-defined heap of water which continued on its course along the channel without change of form or diminution of speed." He followed it on horseback and did so for several miles. This strange wave that he observed in the canal is something that we've actually seen a number of times in physics. In fact, I have an animation here. Normally we think of waves as the nice gentle sort that you see rolling in at the beach, and they come in one after the other. You don't see solitary waves.

The kind of wave that John Scott Russell reported that day was sort of something like this, where you see a single solitary wave as it moves along a channel. Now this kind of wave is very special. Even though it's a wave, it actually has properties of particles. You know, after all, if you take two particles and bump them into each other, they bounce off of each other. Ordinary water waves, if you collide them, actually pass right through each other, but not these objects. If you take these waves, these solitary waves, which we call *solitons*, then in fact, they have the properties of particles, a very strange result. Solitons, in fact, have been observed in many; even to this day, we observe solitons. For example, there are pictures that NASA takes of the ocean where you can actually find single, large, solitary structures that stretch out for hundreds of miles. In fact, they're examples of these solitons. So solitons are a feature of our world. They are waves, but they have the properties of particles.

Remember this mystery of the 496? I told you that Gross and company didn't have 496 anything in their work. Well, the point is that the bosons, or an alternative description called chiral fermions, also called heterotic fermions, can be used to describe solitons. In fact, if you look at the original paper of Gross and company, what they do is say that the charges that make this magical number of 496 are distributed between the ordinary waves of their theories, the things that are described by the chiral bosons or the chiral fermions. These are mathematical words for ordinary waves. Then they need to find the remainder of this 496. It turns out that when you study these ordinary waves, the equations for the ordinary waves in the theory, you find out that if you add up all the ordinary waves and all of the solitons that these waves can describe—these single waves that we've seen in our animation—the number adds up precisely to 496.

So in a sense, the heterotic string, which is the thing perhaps that most forcefully tells us that Einstein's dream has been realized, because the heterotic string is different from the SO(32) and makes it glaringly obvious that unification of the type that Einstein wanted, his so-called unified field theory, is a mathematical reality. In a sense, this successful construction of the heterotic string owes part of its intellectual debt to a Scotsman who was out riding his horse in 1834 and observed a solitary wave and tried to write equations to describe it. The same phenomenon was what was necessary for the success of the heterotic string in Princeton.

Lecture Fifteen
Princeton String Quartet Concerti—Part II

Scope:

The initial work of the Princeton String Quartet resulted in a combination of two strings from different dimensions, a left-moving superstring and the old bosonic right-moving string. But this work did not directly take into account the 496 charges required for anomaly freedom. In this lecture, we'll attempt to find one more description of the heterotic string, based on angles associated with rotations, that helps us arrive at that magic number.

Outline

I. The old bosonic string had 25 spatial dimensions and 1 temporal dimension. If we looked at Einstein's hypotenuse for such a string, the mathematical formula seemed very long and complicated.

   A. Introducing supersymmetry, the form of Einstein's hypotenuse becomes much simpler. The number of lengths is reduced from 25 to 9.

   B. Each one of the lengths in Einstein's hypotenuse is a mathematical attribute found in string theory. Describing the string as having 25 dimensions requires 25 mathematical "objects" to describe it.

   C. In this way, the string is not a single strand. The old bosonic string had 25 plus 1 strands. The newer superstring has 9 plus 1 strands.

II. Last lecture, we learned about the construction of the heterotic string.

   A. The heterotic string combines left-moving, right-moving, and standing-wave modes. In the earlier animation of a standing-wave string, as it oscillates, note that the nodes, indicated by red dots, do not move. (Remember, *node* means a point at rest; *mode* is a way of vibrating.)

   B. In addition to the standing-wave modes, strings have left- and right-moving modes in which the entire string oscillates and rotates. Both the modes and the nodes rotate as a single unit.

C. This behavior is an echo of real physics, known since the 1870s, when Maxwell included the property of polarization in his equations.

D. Of course, the discussion of the 496 charges was included to avoid the charge-conservation anomaly, yet the Princeton String Quartet did not include 496 strands in their papers on the heterotic string.

E. The 9 plus 1 strands apply only to configuration space. To review: In our world, an object can move front to back and requires one strand to describe this movement in its mathematics. An object can move left to right and up to down and requires one such strand to describe each of these motions. Movement through time, from past to future, requires its own strand. At a minimum, to describe our world, four mathematical "strands" are required, collectively called "the string."

F. To arrive at the magical 496 strands, the Princeton String Quartet had to appeal to a mathematical object called a *soliton*. These strange waves actually appear in nature; if two collide, they bounce off each other.

G. If some strands describe space, is it possible to find a strand-like description of the charges in heterotic string theory?
   1. In the original papers of the Princeton String Quartet, there was no sign of such a description. In one of their mathematical descriptions, a set of 16 bosons and 480 solitonic waves appears.
   2. The original papers also introduced a description that used 32 fermionic strands and 464 solitonic waves.
   3. How can these two descriptions relate to the same mathematical object? Remember, bosons carry the forces in our universe, and the objects the bosons act upon are fermions. Bosons and fermions are not the same kinds of objects.

III. The answer to this question is related to some aspects of our world.

A. In 1911, mercury was cooled for the first time to 452° below 0, where it loses all resistance to electrical current and becomes a *superconductor*. As electric current is sent through ordinary wires, part of its energy goes into heating the wire. In a superconducting wire, this does not happen, which has the potential to save a great deal of energy.

B. Superconductors were not fully understood until Bardeen, Cooper, and Schrieffer wrote equations to explain how they work.
   1. In our world, electrical currents are transported by electrons.
   2. The mathematical work on superconductors showed that, sometimes, electrons, even being fermions and obeying the Pauli exclusion principle, can join together as if they were a single object. The resulting object, in many ways, has the properties of a boson, an important fact that allows superconductivity to work.

C. If it were possible to combine an electron and a photon, mathematically, this object would behave like a fermion. If it were possible to join two photons together, the resulting composite would behave like a boson. The mathematics of these combinations is like a multiplication table.
   1. Bosons and fermions in this multiplication table act like the numbers +1 and –1. Thus, +1 × +1 = +1, or boson × boson = boson. Similarly, –1 × +1 = –1, or fermion × boson = fermion, and –1 × –1 = +1, or fermion × fermion = boson.
   2. The behavior of the joint composites of these objects is determined by assigning +1 for bosons and –1 for fermions. These rules apply to any number of fermions or bosons joined to form a composite, e.g., a composite of three fermions acts as a fermion.

D. In our world, it is ultimately possible to determine whether these combined objects are made of two bosons or two fermions, but in the two-dimensional world of the string, this is not possible, because fermions can perfectly disguise themselves as bosons. This explains why the original work of the Princeton String Quartet had two different descriptions for the same theory: In one of the descriptions, the fermions join together to behave as if they are bosons.

IV. The original papers on the heterotic string include strands, but the number of strands associated with electric-like currents never actually equals to 496; it is the strands (not associated with motion in space) plus the solitons that equal to 496. This isn't completely satisfactory.

A. A simple way to find the influence of the heterotic string on all its vibratory modes that have the properties of the denizens is to "count" the amount of energy carried by each strand. Because the original descriptions of the heterotic string don't have all the objects necessary to make that count, there must be another formulation.

B. To get at this other formulation, we again use the example of the ladder and the house. The square sizes are changed by using different orientations of the ladder, starting with it as the base of one square and the top of another square. A third square appears by angling the ladder. No matter how we orient the ladder, the area of the green square is equal to the area of the blue plus the area of the brown.

1. We can write the Pythagorean Theorem to show that the area of the green remained the same, but this equation allows trades of area between the area of the blue and the area of the brown.

2. In the first equation, when the ladder is on the ground, the first area is built using the length of the ladder; because none of the ladder is lifted up on the wall, the second area is 0. When the ladder is in the vertical position, the first area is 0, and the second area, using the length of the ladder which was 0, is now the length of the ladder.

3. In the real world, a person is required to move the ladder. But there is a mathematical object whose role it is to take the ladder from the ground and place it against the wall of the house. These objects are called *generators for symmetry*. Recall that angles measure how much the orientation of the ladder is changed. The *exchanger* (or generator of the symmetry) is a mathematical representation of the person who implemented the exchange!

   4. For the heterotic string, we have 496 generators. (Remember that the heterotic string is an object that acts as if it has 32 different directions in which it can rotate.)
 C. Sophus Lie created the general mathematical tools for describing rotations of related motions. The challenge is to use these tools to provide one more description of the heterotic string, one in which there are exactly 496 strands (in addition to the 9 spatial strands and 1 temporal strand necessary for position in space for the heterotic string).
 D. The standard model contains force carriers, and each of the force carriers is associated with one of these generators. In order to make the heterotic string appear mathematically closer to our world, it must possess strands of the string that mimic gauge parameters (Kemmer angles) associated with gauge invariance of the force carriers.

V. Is it possible to find this last formulation? The answer is remarkably simple.
 A. The angles of the 496 angles associated with the E_8 rotations can be interpreted as being the missing strands. However, in order to write consistent mathematics, we must identify the angular variables with the variables that describe the 496 strands.
 B. The way to solve the missing-strand problem is to introduce 496 angles associated with right-movers of the string. No additional left-movers are required since these are already contained in the standing waves.
   1. In the animation of polarization of light, if the right-moving circle is added to the left-moving circle, one obtains exactly the circle in the middle—the depiction of linear polarization.
   2. The standing-wave modes of the heterotic string actually contain both left and right modes hidden within them, in the same way that the graphic of polarization contains both left-circular and right-circular modes in the linear mode.
   3. To consider the full version of the heterotic string, a series of additions is required. First, 10 of the middle

pictures from the polarization graphic are needed, because those describe the 10-dimensional structure of the heterotic string. Next, 10 left-movers are required to provide the mathematical description of the spin of the heterotic string. Finally, 496 right-movers are needed to describe all of its charges.

    4. This is the mathematical structure of the heterotic string—the goal toward which this discussion has been leading; it includes the angles that look just like the angles associated with change of force carriers in the standard model.

VI. In our last lecture, we talked about the fact that photons are not directly measurable; it is the *change* in photons that generates electric and magnetic fields, and that is what can be measured.

  A. We also said that two different photons could produce the same change, or the same electric and magnetic fields. What's the relationship between these two photons? They are related to each other by an angular-like variable; therefore, charge conservation must also be related to the appearance of such angle-like variables.

  B. In our world, in discussions about the forces for electromagnetism, extra angles do not appear; we see the force carriers that are related to each other by fields we can measure. The ambiguity in the relationship between the "measurables" and the mathematical descriptions is what results in these angle-like quantities.

    1. For the eight gluons in our world, there are eight angle-like quantities associated with gluons. Similarly, in our world, possessing one photon, there is one angle-like variable associated with the photon. Finally, there are three force carriers for the weak nuclear force; therefore, there are three angle-like variables associated with these force carriers.

    2. These angle-like variables actually occur in our world, but they don't refer to directions that we see. They refer to the fact that we are talking about a change in a photon or a photon-like object.

  C. The now-solved puzzle of finding the last formulation of the heterotic string was one in which these angle-like variables could appear. This last formulation is my own research, conducted with Warren Siegel.

D. Let's summarize the three formulations of the heterotic string:
1. The original formulation of the Princeton String Quartet used fermions, the −1 variables. The second formulation used bosons, the +1 variables. But neither of these formulations included the 496 angles that make the heterotic string look like our world.
2. Only in the third formulation can we clearly see that the exact same mathematical structures that describe the standard model are present in the mathematical structure of string theory. This gives us more confidence that string theory is connected to our world.

Readings:

Greene, *The Elegant Universe* and *The Fabric of the Cosmos: Space, Time and the Texture of Reality*.

Kaku, *Hyperspace: A Scientific Odyssey through Parallel Universes, Time Warps, and the 10th Dimension*.

Hoddeson and Daitch, *True Genius: The Life and Science of John Bardeen*.

Zee, *Fearful Symmetry: The Search for Beauty in Modern Physics*.

Questions to Consider:

1. What was missing in the first two formulations of the heterotic string that prompted the search for the third?
2. Summarize the three formulations of heterotic string theory.

Lecture Fifteen—Transcript
Princeton String Quartet Concerti—Part II

In a previous lecture, we talked about the fact that strings are very peculiar in many ways, and it's mathematics that we're doing. Throughout most of these lectures, however, I've endeavored to remove the mathematical language and instead to talk about the mathematics in terms of the pictures that sit behind it. There are some times, of course, when it's good to go back to the mathematics because there're some essential points to make. The transparency I have on the screen here is one reason for doing this. The old bosonic string had 25 spatial directions and one temporal direction; so that if we talked about the Einstein hypotenuse for such a string, it's this awful formula that you see here. Every one of these L's is a different direction. So L_1 would be the first direction, the second direction, until you get all the way down to the 25th direction.

Now the other thing that had happened in string theory in our previous discussions is that we've learned that when you introduce super symmetry, the form of the Einstein hypotenuse changes and becomes simpler. In fact, instead of having this expression going from L_1 to 25, we would only need it to go from L_1 to L_9, not L_{25}. Now I have to reveal something to you that I've kept secret before now. You see, each one of these L's really is a mathematical entity that we find in string theory. So when we talk about the string being 25 dimensions, in fact, there are 25 mathematical objects that you need to talk about the string. Now the important thing of this observation is it means that a string is really more like a string instrument. It's not like a single strand at all. In fact, the old bosonic string should be thought about as kind of a guitar, some sort of a guitar or violin or whatever your favorite string instrument is.

The old bosonic string has 25 (+1) strands. The newer version, the superstring, has 9 (+1) strands attached to it. So that's the only use for this, that we have to recognize that strings, although we talk about a string as if it's a single object, it's actually more like a stringed instrument, very much like a musical instrument. Now we, in our last lecture, followed the Princeton String Quartet with the construction of the heterotic string. You'll remember the heterotic string is this interesting object that combines left movers, right movers and standing wave modes. Let's look at their pictures once more. So here we have a standing wave mode. We'll watch it oscillate. As you can see, the red dots in this diagram do not move. The red dots are what we mean by the nodes of the vibration.

On the other hand, you can see there are these nodes that periodically disappear and reappear in the strings. The number of objects that we see there constitutes the modes of vibration. So you must distinguish the word node from mode. Basically, node means a point that's at rest; mode means a way of vibrating and, of course, the two are connected because you can't have nodes without modes. On the other hand, in addition to the standing wave modes, we have the left-moving modes. Here's a left-moving mode. As you can see, it's actually—there's a previous picture, but where the entire string rotates around. So therefore, the modes are rotating, and the nodes are rotating. They have to move as a single unit. This is a picture of a left-mover; then, of course, there are also right movers. As you can see, the only difference is that the modes rotate in the opposite sense.

In our last lecture, we also discussed how this was an echo of real physics that's been known since the 1870s. When Maxwell wrote his equations about light, it was observed that there was a property of light called polarization, which essentially told us what was going on with the electrical field associated with a particular light ray. Now we've also solved the puzzle in our last lecture. We said that you needed to have 496 magical charges in order to avoid the anomaly, and when Gross and company wrote their paper on the heterotic strings—the first two papers, at least—you will, in fact, not find 496 strands, because that's the important point to note. The strands that we talked about previously—which only applied to configuration space—let's review this idea. In our world, I can move frontward and backward. That's one independent direction. You would need one strand to describe that in the mathematics. We can move left to right. That's a different direction. In terms of mathematics, we would need an object to describe that. Then finally we can move up and down. That would be a third object. Anything else we do is a combination of those three, so three strands would suffice to describe our world.

But of course, we need time, also. So at a minimum, to describe out world, a string would consist of four mathematical objects that we would call the string, but you can think of them as the strands of the string. It turns out that in order to get this magical 496, Gross and company had to appeal to a very strange mathematical object called a soliton. I explained in the last lecture that there are these strange waves that have been observed in nature, can be observed, and they have the property that these are waves, which actually bounce off of each other just like real particles. Now this is a very strange way to describe the theory, because you might say well, gee, if you have these strands for describing space, is it possible to look for this strand-like description of the charges in the theory?

Now in the original papers of Gross and company, you will find no such signs of such a description. In one of their mathematical descriptions, they have a set of objects that are bosons. Again, I remind you bosons are things that spin at even numbers times the spin of the electron. They have 16 bosonic strands in their original papers, and then they count 480 of these solitonic waves. The other thing that their original paper does is introduce 32 fermionic strands, then they count the solitons associated with that, and they turn out to be 464. So the sums of these numbers are always adding up to the magic 496, which we need in order to get the charge conservation and the absence of the anomaly. There's something here about which you ought to stop me. How can it be that these two things describe the same mathematical object? After all, I told you in our world, things such as bosons are the force carriers. The particle of light is the boson, the gravity. The graviton is a boson. The intermediate vector bosons—they even have the word boson in their name—are bosons. These are things that carry the forces in our universe.

On the other hand, the objects that the forces act upon, such as the electrons, are fermions. They all obey the Pauli Exclusion Principle, which is the basis for chemistry. So when I tell you this, you ought to stop me, because you should say, "Wait a minute. Bosons and fermions are not the same objects. They're not the same kinds of objects. So how can you use them to describe a single mathematical entity?" Well, the answer to that turns out to be related to some aspects of our world. And I've endeavored throughout these lectures to make sure that, although we talk about the mathematics of string theory because it is a theory of mathematics, we always try to make contact with the mathematics that is used to describe things in our world. So we're going to do this one more time.

In 1911, mercury was cooled for the first time to 452 degrees below zero. When that happens, mercury loses all resistance to electrical current. You can send a current through mercury cooled to that level, and it flows with no loss of energy. That's very different from our world. Now when that occurs, we say that the material has become a superconductor, and superconductors are among the Holy Grails of our technology. Because you see, ordinary wires that we use today, as you send a current through the wire, part of the energy of that current goes to heating the wire. Well, that's inefficient because it means that all the wires that go from the generator stations to your house are eating up energy that you can't use in your house. In fact, that energy shows up in heat, and we can't do anything with that heat.

So if we had a superconducting wire, it means that we could actually save energy because, instead of wasting some of the energy in the wires, it would all go to the households to be used. So that's why superconductors are so important.

Superconductors, although they were first discovered in 1911, weren't really understood until 1957. Three physicists, Bardeen, Cooper and Shrieffer, wrote equations to explain how superconductors work. In these equations, there's a secret. In our world, charge is transported in electricity because electrons carry the charge back and forth. What these mathematical works about superconductors showed was that sometimes electrons, even though they're fermions and therefore obey the Pauli Exclusion Principle, sometimes they can join together as if they're a single object. Now this single object that you get out of the composite, in fact, has the properties of a boson, and this turns out be a very important fact that allows superconductivity to work.

So you can now ask the question—well, that's interesting—we take an electron and an electron, join them together in some manner, and they then behave like a boson. What would happen if I took an electron and somehow magically joined it to a photon? What would that object behave like? Well, the answer turns out to be that it behaves like a fermion. In fact, there's one final you could ask. Well, suppose I join two photons together? What does that composite behave as? And the answer is that it behaves as a boson. In fact, the statements we've just made, we can put in the form of a little multiplication table, and here it is. It says that if you take two bosons and join them together, you have something that behaves like a boson. If you take a boson and a fermion and join them together, you have something that behaves like a fermion. And if you take two fermions and join them together, you again have an object that behaves like a boson.

Now you might say well, gee, who cares about all that? But it has important implications. If you actually look at this table, you'll notice something else. Bosons and fermions in this little table act exactly like the numbers + and -1. After all, +1 times +1 is +1. Well, boson times boson is equal to boson. Boson times fermion, well, that's like -1 times +1, but -1 times +1 is -1. Well, gee, that's a fermion. Similarly, if you take two minus signs and multiple them together, you get a plus sign, so that's a boson. So the message here is that when you take joint products of these objects, you can actually figure out their behavior by simply assigning the number of +1 for every boson and the number -1 for every fermion.

Now in our world, even when electrons join themselves like this together, at the end of the day, we can tell something—if it's made of two bosons that are put together or two fermions that are put together. But in the world of the string, remember the string is effectively a two-dimensional world because the string has an extent, and there's a time at which you look at it, so that's two dimensions. In the world of the string, however, it turns out that this multiplication table that we just went through is exact. In a two-dimensional world, fermions can perfectly disguise themselves as bosons, and this explains the mystery for why the original work of Gross and company has two different descriptions that apparently describe the same theory. Because in one of them, what you have is that the fermions are joining together to behave as if they are bosons.

There's something else about this 496 that we can talk about because we haven't actually seen it. If you look at the original papers on the heterotic string, we have the strands, but we never have the strands actually equal to 496. We have the strands plus the solitons equal to 496. Now there's a reason why this is not completely satisfactory, because if you're asking the question what is the energy of the string, the easiest way to do that is to count the amount of energy that's carried by each strand. Since the original descriptions of the heterotic string don't have all of the objects that are necessary—remember, we had to count solitons to get the charges right—that suggests that there's another formulation of this original work in 1984. To get at this other formulation, once again I'm going to turn to the example of my ladder. I know you're probably tired of it by now, but it contains so much essential physics in terms of understanding the issue of symmetry. That's what we keep returning to, the issue of symmetry.

So here's our return to our favorite house, and there once again is our ladder lying on the ground. We can, of course, again bring ourselves close by it and grow our squares so that there's one square with an area. This ladder, for this one, is the base of the square. But we can grow a second square downward where the ladder acts as the top of the square, and since these are both squares, the areas are the same. Now as we tilt the ladder up, we begin to learn about invariants one more time. The area of the green is equal to the area of the brown plus the area of the blue, and this is true no matter how we orient the ladder. In fact, we can put the ladder totally against the wall, in which case, it's the brown and green areas that are the same. So why do we need that in our discussion here?

Well, in the context of the ladder that we just saw here, where there was something that was the same, namely, the area of the green, but we could trade it back and forth between the brown and the blue. You could imagine writing some equations that show that, and I've done that on this transparency. When the ladder is directly on the ground, we can say well, gee, there's one number that's the total length of the ladder, and since none of it is up against the wall, the other number is zero. On the other hand, when we have our ladder totally vertical, then the first number that used to be the entire length of the wall is zero, and the second number, which used to be zero, is now the entire length of the ladder. So that's the mathematics. Let's look at the two pictures.

Here is the second length. Remember the second number talks about how much of the ladder is up against the wall. Well, how much of the ladder is against the wall is exactly its length in this picture. Whereas, when you go all the way back to the first picture, it's actually the length along the ground that's the entire length of the ladder. That's typically the way symmetries act. So what turns the ladder? Well, if I had a real ladder, I would have to do it. I'd have to take the ladder, have it lying on the ground and lift it up until it's standing vertically against the wall. Now there's a mathematical description of that process, and it involves describing an object whose role is to exactly take the ladder from its position on the ground, to putting it up against the wall. We've met these objects before in our lectures, and they're called generators for symmetry. So we need a generator here.

Now in the heterotic string, we have our 496 generators. Those are the things that—remember, the heterotic string acts as though it's an object that has 32 different directions in which you can rotate—that are where the number 496 comes from. So we want to try to get the 496 in a more explicit way. Now, Sophus Lie, the mathematician who first invented the kinds of descriptions of these things, gave us the tools for describing rotations, and in particular, we want to use his tools to provide one more description of the heterotic string. This final description, as I mentioned earlier, will allow us to account for energy. In fact, what we're looking for is a description where there are exactly 496 strands in addition to the ones that are necessary for space. Remember, we need nine spatial and one temporal strand to talk about position in space for the heterotic string. But we also want a theory where there are 496 strands that allow us to talk about the energy in a way that's completely consistent.

There's another reason why we want to find this formulation. Remember when we talked about the Standard Model. In the Standard Model, there were the force carriers, and each of the force carriers is also associated with one of these generators; that is, one of these things that effectuates a rotation. So in order to make the heterotic string appear mathematically closer to the Standard Model, you would actually want to find the angles that are associated with the turnings that are present in the heterotic string. Now this was not in the original work of 1985/1984 for the heterotic string, but you know, in physics, we keep trying. So how do we find this last formulation? The answer turns out to be remarkably simple. The angles that I talked to you about earlier, well, those angles, one can interpret themselves as being the missing strands that we're looking for.

However, in order to write the mathematics in such a way that's consistent, what we have to do, which is quite remarkable, is identify the angular variables that we have been speaking about with the variables that describe the 496 strands. So in particular, the way to solve our missing strand problem is to look at the right movers, not introduce a single right mover; imagine 496 right movers in this picture. That would be the missing angles that we need. Now what about the left-moving part of the heterotic string? Well, in the left-moving part, it turns out that we don't actually need any additional left movers if we compare to the original paper of Gross. All the left movers in the original paper of Gross and company are, in some sense, contained in the standing waves.

You might say well, gee, wait a minute. You keep saying left, right and standing. Well, let me show you a secret to that. Perhaps you'll remember in our discussion of light, I've shown this picture before, and I said that the middle picture talks about a light ray where the electrical field simply moves up and down. If we put this in motion, there it goes, the electrical field vibrating up and down. There's actually something quite remarkable about this picture, which is that if one were to take the right mover—so let's watch right movers in the upper right-hand corner—you'll notice that it rotates around in a right-handed sense. Compare that to the left mover, which rotates around in a left-handed sense, and now let's do something remarkable. Let's say suppose we were to take the two top pictures and put them one right on top of the other. What would be the result? Well, the answer turns out to be that if you take the picture in the upper left-hand corner and add it to the picture in the upper right-hand corner, you get exactly the picture in the middle.

How can that be? Well, let's see if we can figure this out. You see, the upper left-hand picture here has a green arrow, which is oriented exactly in the same direction as this one. If we were to put these two pictures together, this green arrow would sit right on top of that one. Therefore, if we tried to figure out the—we could slide it along, and we'd get a total length that's twice the length of a single arrow. Well, if you look very carefully in our middle blue line, it's actually twice the length. So now let's move the two arrows until they're pointed like this. Now you'll notice there's something interesting. The two arrows, which you can think about as two forces pushing against each other, are exactly canceling each other out because they're equally large. Well, in the middle picture, we see there's nothing there. They have canceled each other out, and finally, rotating the arrows, continuing, we see that when they point down, the middle picture points down exactly.

There's a lesson in this. The lesson is that in the heterotic string, if you look at the standing wave modes—they actually contain both left and right modes hidden within them—when you consider these 496 strands that we're talking about, the way that the picture works is as follows. You simply do a bunch of additions. You need ten of these middle pictures because those describe the structure of the ten-dimensional world in which the heterotic string lives. You actually need also, curiously enough, ten left movers. We can consider those left movers, and you need those for the heterotic string, but not to describe motion in space. Remember, the heterotic string is supposed to spin. Well, you need the mathematical description to provide that spin. That's what these objects are, and finally, we need these mysterious magical 496 charges. That's what these are.

So the heterotic string, if you want to think of it this way, is you take ten times the middle picture here, you take ten times the picture on the left, and you take 496 times the picture on the right. That's the mathematical structure of the heterotic string that we're looking for. This has the angles that look just like the angles associated with changing things in the Standard Model. You might say, "Wait, Professor Gates, you didn't talk about angles in the Standard Model," but you see, I did. In our last lecture, we talked about the fact that photons are actually not directly measurable. It's the change in the photons that causes the electric and magnetic fields, and that's what we can measure in our world. If that's the case, you will also recall that I talked about that you could have two different photons, but whose changes, even though they're different, the change in each of them is the same; therefore, two photons that are totally different are able to produce exactly the same electric and magnetic fields.

So what's the relationship between these two photons that produce the same electrical field? The answer turns out to be that they're related to each other by an angular variable; therefore, this mysterious thing that leads to charge conservation is also related to the appearances of a certain kind of angle. Now in our world, when we talk about the forces for electromagnetism, we don't actually see any extra angles. We just see the force carriers that are related one to another by fields that we can measure, things that are the electric and magnetic fields. This ambiguity in the relationship between the measurables and the mathematical descriptions is what gives us these angles. So in our world, we have the gluons—there are eight of them. That means that there are eight angles actually associated with gluons. In our world, we have one photon. That means there's one angular variable that's associated with the photon, and finally, in our world, there are three force carriers associated with a weak nuclear force; therefore, there are three angles.

So these angular-type variables actually occur in our world, but they don't refer to directions that we normally see. They refer to the fact that we're talking about the change in the photon or the change in a photon-like object. So we've actually solved the puzzle here in terms of finding this last formulation of the heterotic string where these angular-like variables appear. Now I must admit I'm really happy to have the opportunity to talk to you about this last formulation because, you see, this last formulation is my own research. Although the heterotic string was constructed in 1984/85, this final formulation, where we only found all of the angular variables associated with the string, didn't occur until 1988. This was in work that I carried out with a colleague; in fact, his name is Warren Siegel, and this final formation has all the angles. So I'm going to give you three scores for the heterotic string now.

In the original work of Gross and company, there's what I call Formulation A. You can think of this as a score. I've used some very sophisticated mathematics to shrink it down to this simple expression. So this is the formulation where they use what we called fermions. Those are the -1 variables that we talked about at the beginning of the talk. There's a second formulation where they use the bosons. These are all the +1's in our discussion of bound states. But neither of these formulations had the 496 angles that make the heterotic string look like our world. There's a third formulation, which is the work here, which is my own research with my colleague, Warren Siegel, and it is only in this third formulation that you clearly see that the exact same mathematical structures which describe the Standard Model in our world are precisely present in the mathematical discussion of string theory.

This gives us confidence that string theory is really connected with our world. One of the things that this last piece of work does that's different from the others is it allows us to measure energies in a much more precise way for all of the vibrations that come from the string. In particular, there are lots of vibrations that are called spin-zero vibrations, that are very difficult to see in other ways of writing the thing, that are very apparent in this last one. So for me, the lesson is that you get to have fun in physics where you can make a contribution by following these very advanced topics, and you have a very great sense of satisfaction that perhaps, just perhaps, you've added a single nugget, a single strand, in the tapestry that will be physics.

If we're lucky, perhaps one day, perhaps in a few hundred years, maybe, this kind of mathematics will be important for constructing some useful device or process in our world. After all, that's the purpose of theoretical physics. We pursue it at one stage where it's mathematics. If we're lucky and have it right, at the final stage, it becomes something important in our world. We saw this in the story of Maxwell. He predicts that electromagnetic radiation from his mathematics is light, and we have cell phones as a consequence.

Lecture Sixteen
Extra Dimensions—Ether-like or Quark-like?

Scope:

After Maxwell, the scientific establishment believed that light, as a wave, propagated through a medium called the *luminous aether*, or *ether*. When scientists tried to measure the velocity of the Earth relative to the ether, they got null results. With Einstein, it was instead understood that time and space measured by different observers in different frames of reference yield different results, but it is only the relationship between the numbers that changes, not the object being measured itself. In string theory, the mathematics of the heterotic string can be interpreted in two ways. The most popular interpretation argues for extra dimensions. The alternative asserts that what may be thought of as angles arising from extra dimensions are actually angular (or angle-like) variables associated with the change of force-carrying particles. This interpretation lends itself to a much more flexible interpretation of string theory.

Outline

I. Before Einstein, the scientific establishment believed light, like all other known waves, propagated in a medium called the *luminous aether*, or *ether*.

   A. With special relativity, we realized that the way in which space and time is measured depends on how this is done *relative* to some motion. An important consequence of Einstein's idea was that even though measurements of an object may appear different to different observers, the object itself is not actually changing. The properties of an object that remain unchanged are described using Einstein's hypotenuse.

   1. A line segment, stretched in one direction, could be measured with a yardstick to be 3 feet long. If the yardstick was oriented perpendicular to the line, the measurement would be 0. The two numbers (3,0) describe the point at the end of the segment. If the line segment were re-oriented 90° with respect to its original direction, the two measurements could change to 0,3.

      2. Notice that nothing happened to the line segment; only the mathematical descriptions of the position of its tip changed.
- **B.** Einstein realized that duration and length measured in one frame of reference can differ from the same measurements made in a different frame of reference. But nothing about the object being measured has changed!
- **C.** The mathematical description for trading time and space measured by one observer with time and space as measured by another observer is given by a set of formulas called the *Lorentz transformations*.
- **D.** In string theory, there is the idea of extra dimensions, but there are no experiments yet that *demand* the scientific belief that these extra dimensions actually exist.
- **E.** In 1999, two physicists, Randall and Sundrum, proposed a view that our universe may be described in a larger world as if it were a pane of glass. This is called the *brane-world scenario*. With acceptance of brane-world scenarios, certain things in physics become much more obvious.
  - **1.** In our world, the gravitational force is the weakest of the forces that act on matter. Why is gravity so weak?
  - **2.** If our world is like a pane of glass, perhaps some gravity is "leaking" into other dimensions, thus diminishing its effects.
- **F.** Though widely accepted by the majority of scientists who consider such matters at present, the *brane-world scenario* cannot be settled until there is some sort of experimental evidence, either direct or indirect.

II. The idea of "extra" dimensions did not start in string theory.
- **A.** In the 1930s, Kaluza proposed that the physics of our world required at least one extra dimension. When Einstein first read the paper, he thought it was nonsense, but on later reflection, he found it very interesting to consider: Maxwell's equations could be found in the mathematical results combining Einstein's own theory of general relativity with the idea of the extra dimension.
- **B.** Klein pointed out that if this extra dimension were very small, it would be easy to overlook it.

1. Imagine a house with many rooms, but one special room possessing a doorway so small no one could enter the room. If one were able to pass through the doorway, it would be like any other room; it's the door that's the problem.
2. A way to avoid the problems of extra dimensions is to think about them as being very small or possessing very small "doorways."

C. In 1978, a group of physicists studying supergravity realized that the equations for the most complicated version of this theory could be simplified by the introduction of 11 dimensions. Thus, when string theory had its breakthrough in the 1980s, physicists were already thinking about extra dimensions because of supergravity.

D. Are extra dimensions real? Present-day technology is limited in its ability to answer this question. This situation often occurs in physics.

III. The idea of extra dimensions has also shown up in another way in physics.

A. Earlier, it was thought that the smallest electrical charge was the size of that charge carried by the electron. With the acceptance of the existence of quarks, evidence showed that electrical charge could come in fractions of 1/3 or 2/3 the charge of an electron.

B. Charge still appears to occur only in "chunks," but these chunks are of a smaller size than originally thought. Physicists have long wondered why this is so. One of the ways to answer this question was proposed by Kemmer in 1938 in his explorations of protons and neutrons.
1. Accounting for spin, electrons seem to come in only two varieties; they either spin up or spin down. That is similar to electrical charge. Kemmer tried to apply the mathematics of spin to charge.
2. Protons are charged; neutrons are neutral; and the two are almost the same in terms of their mass. Kemmer posited that, on some deeper level of reality, protons and neutrons had the same mass. If that were true, then it is only the charge that distinguishes them.

3. He further posited that charge was the same as spin. If that were true, protons and neutrons would not be different objects but the same object "spinning" in different ways. The implications of the mathematics of this hypothesis, known as *isotopic charge space*, work perfectly well to describe the physics of the denizens. The mathematics of spinning objects in our world can be applied to things that possess electrical charge.

   4. Kemmer's notion of charge also includes mathematical objects that behave like angles, but they are not dimensions to which we have access. In fact, his angles are precisely the ones associated with the force carriers. Ordinary angles describe different orientations of objects in space. So the occurrence of large numbers of angles would seem to imply lots of different directions for space.

C. Throughout, the standard model uses the idea of mathematical quantities with the properties of angles. Recall the fact that changes in two different photons can produce the same electrical field. Two photons that possess this property have mathematical descriptions that are related to one another by an angle.

D. Even for physicists who do not accept the belief in extra dimensions—and many physicists don't—the mathematics that describes the electromagnetic force behaves in many ways as if it has three spatial dimensions, one temporal dimension, and one other dimension that allows for the angle described above.

IV. Now a return to one of the "dirty little secrets" of the heterotic string.

A. Recall diagrams representing the polarization of light—also related to descriptions of standing waves, right-moving waves, and left-moving waves in the heterotic string.

B. The standing-wave mode corresponds to one of the strands of the string that describes space; thus, we need nine of these standing-wave modes to describe the space of the heterotic string and one to describe time.

C. When the animation is in motion, we can see that superimposing the representations of the left- and right-movers actually yields only the standing wave. Can we reverse that process? Is it possible to mathematically dissect a standing-wave mode into its left- and right-movers?

D. Using a mathematical tool called *Fourier analysis*, we find that the standing wave is exactly equal to the mathematical expression for a left-mover added to the mathematical expression for a right-mover.

E. This means that anything associated with a standing wave can be pulled apart mathematically into left-movers and right-movers. Why is that important? It turns out that the angular variables of string theory, which are associated with the different photons that produce the same electromagnetic force, are always either purely left-movers or purely right-movers, but never standing waves.

F. This understanding gives string theory a kind of flexibility. What might be initially identified as a dimension (associated with standing modes) can actually be pulled apart and traded for more charge! The mathematical structure of the model can be reduced to less than 10 dimensions. Effectively, a trade-off of directions for charge occurs.

G. This is a potent idea. In studying the mathematics of the heterotic string, there can be two interpretations. One (the presently most popular) interpretation argues for extra dimensions, but the alternative asserts that what can be misidentified as extra dimensions are actually angular variables associated with the changes of force-carrying particles.

H. In string theory, what has traditionally been identified as a dimension may actually be two kinds of charges: left charge and right charge. In the original heterotic string, all 496 charges are associated with right-movers and the theory possessed nine spatial dimensions and one temporal dimension.

    1. It is possible to carry out an operation on one of those nine mathematical objects describing the space of the heterotic string and pull it apart into *left-charges* and *right-charges*. In fact, if this is done to six of these dimensions, the result is an object that exists in four dimensions (which looks a lot like our world).

    2. The mathematics of the heterotic string allows a construction where the number of dimensions is reduced to four, but in the process, it is necessary to introduce more than 496 right-charges.

V. One might also wonder what is happening to the left-movers. For every extra dimension encountered, after being pulled apart to get a right piece, which is a charge, there remains a left piece. With what should it be identified?

    A. My colleague Warren Siegel and I pursued this question of mathematically pulling the heterotic string apart and found another formulation of the heterotic string that has the same 496 currents as in the original version and something else.

    B. The discussion about the denizens of the quantum world included the fact that the electron possessed copies. One copy of the electron, called the *muon*, is 200 times as heavy, and a second, the *tau particle*, is 1700 times as heavy.

    C. For a string theory in four dimensions, this work showed that the mathematics automatically produces copies of particles in the theory. Thus, a string theory written in four dimensions looks more like our world than a string theory written in higher dimensions.

    D. Without string theory, there is no explanation for why copies of the electron exist. With string theory, we can see that the copies correspond to the absence of extra dimensions, which were pulled apart into charge and a property that is responsible for what is called *family number* (i.e., the copying) in the standard model.

VI. Most of the popular discussions about string theory include the process of compactification.

    A. Previous discussions considered that extra dimensions might exist and be very small, but they might also be incredibly warped. If string theory allows for extra dimensions, it ought to explain why they must be small and warped. There is no complete and consistent explanation about this known to your lecturer.

    B. On the other hand, if we start from the point of view that string theory ought to describe our world, then by definition, one ought to allow for string theory to naturally provide for the copying processes seen in the standard model, as well as the angular variables that are not associated with actual motion but are associated with objects like the photon. Taking this view, then, string theory comes to live comfortably in four dimensions.

C. In the four-dimensional point of view, there is no question about warping, because the theory has no options. If one believes in extra dimensions, then hidden questions arise: Why did the theory warp itself? Are all possible compactification techniques known?

D. These questions will probably not be settled for decades or perhaps even centuries. Is string theory a four-dimensional construct, or is it a higher-dimensional construct that involves compactification?

VII. Let me make a few closing comments on four-dimensional string research.

A. In the mid- to late 1980s, three groups of physicists studied the question of four-dimensional strings. One group extensively studied the idea that the heterotic string is an actual four-dimensional construct, using a mathematical approach called *free fermions*. Another group used a technique called the *covariant lattice*. The drawback with all these approaches is that strings are much more complicated to write mathematically as four-dimensional objects than their higher-dimensional analogs.

B. In recent years, physicists have been looking at *G2 compactifications*, *spin-bundle compactifications*, and *fluxes*. These are all different mathematical tools for imagining the warping of higher dimensions. But we have no guarantees yet that, with the higher-dimensional point of view, a complete determination of the right way to "bend" the extra dimensions to produce our world has been found.

C. On the other hand, believing in the four-dimensional point of view requires the addition of many more "pieces" to our world that are not included in the standard model. The implications for this are not quite understood yet. If this problem is ever solved, it will likely become possible to write many versions of the string that contain our world and other "stuff." What is this other "stuff"? Perhaps it is the dark matter discussed in Lecture One, copiously produced in a four-dimensional string.

Readings:

Greene, *The Elegant Universe* and *The Fabric of the Cosmos: Space, Time and the Texture of Reality*.

Kaku, *Hyperspace: A Scientific Odyssey through Parallel Universes, Time Warps, and the 10th Dimension*.

Krauss, *Hiding in the Mirror: The Mysterious Allure of Extra Dimensions, from Plato to String Theory and Beyond*.

Halpern, *The Great Beyond: Higher Dimensions, Parallel Universes, and the Extraordinary Search for a Theory of Everything*.

Randall, *Warped Passages: Unraveling the Mysteries of the Universe's Hidden Dimensions*.

Penrose, *The Road to Reality: A Complete Guide to the Laws of the Universe*.

Susskind, *The Cosmic Landscape: String Theory and the Illusion of Intelligent Design*.

Questions to Consider:

1. Describe the two interpretations of the heterotic string discussed in this lecture.
2. What problems are left in the higher-dimensional view of string theory?

Lecture Sixteen—Transcript
Extra Dimensions—Ether-like or Quark-like?

If you looked at the last lecture, there was something you might have caught in my discussion about the angles in the final formulation of the heterotic string. This something has to do with what exactly is a dimension? In string theory, it is often said that strings require extra dimensions. That, in fact, is not quite true, and in this lecture, we're going to talk about that. In science, of course, ideas come into fashion and, with the use of experiments, go out of fashion. Prior to Einstein, the scientific establishment believed that light was similar to all other known waves. In particular, all other known waves must have a medium in which they propagate. Water waves, after all, travel on water. Sound waves propagate through the air. Since there's no air in outer space, there's no sound. Since Maxwell's Equations describe light as a wave, even Maxwell thought there had to be something that supported these waves. This something was called the *luminiferous aether*, or *aether* for short.

The previous lectures have also covered the discovery of Einstein in the form of Special Relativity. In Special Relativity, Einstein tells us that the way that one measures time and space actually depends on the kind of motion one is performing. But an important consequence of Einstein's idea is that, although things appear to be different to a different observer, that doesn't mean something is actually happening to an object. What do we mean by that? Well, in order to illustrate, let's use my arm. If I lay my arm directly out in this direction, you could come along with a yardstick and measure its length. Let's say you've have three feet. If you laid the yardstick in this direction, you would have zero because my arm is not pointed in that direction. So you could write two numbers, three and zero.

On the other hand, if I bring my arm to this position, the numbers that you write are very different. You would write zero for that direction because now the yardstick doesn't measure anything in that direction, but you would measure three feet along this direction. Nothing actually happened to my arm as I swung it back and forth. It's the same arm. It's just that the two mathematical descriptions are different. The important thing that Einstein realized was that this is what happens to time and space that the time you measure if you're in one frame of reference and the length you measure in that frame of reference can differ from the measures made by someone else in a different frame, but that doesn't mean that something actually happens to the object. That's the important message.

In fact, the mathematical description of how you trade the time that's measured by one observer and the space as observed, measured by one observer, and compare those to the time and space measurements of another observer are given mathematically by a formula called the Lorentz Transformation. You notice it's not called the Einstein Transformation. The reason is because it was actually written down by another physicist by the name of Lorentz. You see, Lorentz also believed in the aether. However, when people did experiments to try to measure the velocity of the earth relative to the aether, they had no results. There was no obvious way to measure the aether. One way to explain this result is to say that the devices with which you were making the measurements were actually shrinking and, therefore, that's why you didn't find the measurement that you were looking for.

That's the distinction with Einstein's view that there was nothing going on with the object. It's the relationship between the numbers that are changing and the previous point of view about the aether. So the aether has gone out of fashion in physics. On the other hand, the idea of quarks, as we've talked about in this presentation, is very much in fashion in physics. We have lots and lots of experimental evidence that quarks exist, even though no one has ever isolated a single quark in the laboratory. Some ideas we believe in, others we don't. How do we distinguish? Well, as I tell freshmen in my physics classes, physics is not what we write on the blackboard. It's what is happening in the laboratory. If you find the evidence of the quarks, we are forced to believe them.

In string theory, we have this idea of extra dimensions, but we have no experiments yet that force us to believe that they have to be there, even though you will often hear in many discussions of string theory that string theory requires extra dimensions. This statement is technically false. String theories are consistent with extra dimensions; they don't actually require them. This confusion actually turns out to be tied up with something that we talked about in the last lecture. We used the word dimension. So the question is what exactly a dimension is. Well, from the point of view of adding extra dimensions, which we sometimes call space invaders, that's the extra dimension that you hear about often in string theory. There are extra pieces of space that come into existence.

This idea that these things are there is, in fact, part of the standard view of physics as is widely accepted now. In fact, during 1999, two physicists by the names of Randall and Sundrum proposed a view that perhaps our universe is described in a larger world as if it's a pane of glass. This is called a brane, by the way, and these kinds of view of physics are called brane world scenarios. If you believe in brane world scenarios, it turns out that certain things in physics become much more obvious. For example, we've talked about the coupling constants in nature. Here we, in fact, have a plot of them. In our world, the gravitational force is the weakest of the forces that act on matter. Then there is the weak nuclear force, the electromagnetic force, and the strong force. As we go up in this plot, we're looking at stronger and stronger forces.

Why gravity is so weak, is a question people have pondered. One way to understand that is that if our world really is like a pane of glass, then perhaps some of gravity is leaking into the other dimensions, and that's what the brane world scenarios propose. It certainly has something to commend it, and there are other reasons for believing in extra dimensions. However, we still have to come back to the question of whether they are real. Let's talk about extra dimensions and how it actually entered physics, because the idea of extra dimensions did not start with string theory. In fact, we can go back to the 1930s, when a physicist by the name of Kaluza wrote a paper in which he proposed that we need at least one extra dimension. This paper had difficulty being published. In those days, you had to send your papers to a great physicist. He would approve it, and then it would get published. In the case of this paper, it went to Albert Einstein.

On Albert Einstein's first reading of the paper, he thought it was nonsense, but upon reflection, he realized that there was something very interesting about this paper; that is, if you allowed one extra dimension, then you could find Maxwell's Equations among the mathematical results that lie in the extra dimension. But the problem was, of course, no one had ever seen an extra dimension. This was fixed by another physicist named Oscar Klein, who pointed out that if this extra dimension were really very, very small, it would be very easy to overlook it. Often when people hear the words extra dimensions, they think science fiction, but let me give you another way to think of it. Suppose you were in a house that had many rooms, but that you had a door to one special room that was so small that you couldn't get into it. Then effectively, that other room was not there. If you were able to get into it, it would be pretty much like the rest of the house, but the door is the problem.

One of the ways to avoid extra dimensions is to think about them as being very small, so that we can't get into them. We can't shoot atoms into them. We can't shoot light into them. If that's the case, then maybe they're there, but we can't detect them by ordinary means until we find the experiments that are precise enough to exceed the limits imposed by these small dimensions. So the Kaluza-Klein idea was actually started in the 1930s, but then it went out of fashion until the 1970s. In 1978, a group of physicists were studying something called supergravity. We'll come back and talk about that in a later lecture. They realized that the equations for the most complicated version of this theory were much simpler if they could introduce 11 dimensions. Wow, we're hearing the word 11 now. Remember, string theory likes ten dimensions; that is, nine spaces and one time. This construction wanted ten spatial dimensions and one time, so it goes up by one. When string theory actually had its breakthrough in the 1980s, physicists were already talking about extra dimensions because of this previous theory called supergravity.

Are the extra dimensions real? Well, it's easy to miss things in our world because we are limited by technology. When we looked at our denizens, we found out that the electron had a friend called the neutrino. The neutrino, for a long time, was thought to only come in one spin. It could only spin in the right-handed sense, because that's all we ever were able to measure. A few years ago, in an experiment called Super-Kamiokande, it was verified that neutrinos that spin in the other sense actually do exist, but we simply have not had the technology to find them. Here you can go from an idea of something that's not there, and you find suddenly it's there if you'll increase your technology.

This idea about extra dimensions has also shown up in another way. You'll notice that electric charge comes in chunks. In the old days, we used to think that the smallest electrical charge was the size of that which the electron carried. The electron has negative that number; the proton has positive that number. With the realization of the existence of quarks, the scientific evidence showed us that electrical charge could come in a fraction one-third of that size. So a charged quark has two-thirds the charge on an electron, or minus one-third the charge. Charge still comes in chunks; it's a smaller size. Why is that? Well, this is a question that, actually, physicists have worried about for a very long time.

One of the ways to answer that question was proposed by an English physicist by the name of Nicolas Kemmer. In 1938, he made the following observation. We've talked in this series of lectures about the fact that electrons only seem to come in two varieties; either spin up or the left-handed variety, spin down. That's like electrical charge. After all, at that time, people thought electrical charge was either plus the charge on the proton, minus the charge on the proton, or zero. So there were three possibilities instead of two, but still they're discrete. What Kemmer observed was something very interesting. He said let's look at the mathematics of spin and try to apply that to charge. When you do that, you find out it can actually work. He did this in the context of an actual system in which he was thinking on exploring, namely protons and neutrons.

If you look at protons and neutrons, one's charged, the other's neutral. The neutron actually weighs a little bit heavier than a proton, but that's like one part in 1,800, so they're almost the same in terms of their mass. So what Kemmer said was the following. He said suppose it were the case that, in some deeper level reality, protons and neutrons have the same mass. If that's the case, it's only the charge that distinguishes them. Then he had a brilliant proposal. He said suppose that charge is like "spin", so that, in fact, protons and neutrons are not different objects, but the same objects spinning in different ways; here we put the word spin in quotes. It turns out the mathematics works exactly fine. This was borne out by the idea of isotopic charge space. Now that's an imaginary space. It's not a space in which we can move, but it means that the mathematics of spinning objects that we're used to in our world can actually be applied to things that have electrical charge. So Kemmer's notion of charge says that there are things that are like angles, which mathematically behave as angles, but they're not dimensions that we have access to.

This idea that you can have mathematical quantities that are important for physics, that have the properties of dimensions, but yet are not, is all over the Standard Model. We talked about this earlier. It's, in fact, this business that if two photons have a change that produces the same electrical field then they're related to each other by something, some mathematical object. This mathematical object, it turns out, acts like a dimension. So even if you don't believe in extra dimensions, and there are many physicists who don't, it turns out that when they write their mathematics for describing the electromagnetic force, that mathematics behaves in many ways as if the world in which we live has the usual three space and one time dimensions, and then one other dimension, which is described by an angle. But this angle, of course, is actually the business about two photons producing the same electrical field.

The notion of dimension is not at all as simple as you might imagine from your everyday experience or previous exposure to the discussion of string theory. In fact, it even gets more confusing because we have touched on the aspect of the heterotic string. Now let me tell you one of the dirty little secrets of the heterotic string. We've seen this diagram before, I've used this. I call this the lariat diagram, and we use it in many, many ways in order to describe polarization of light and in order to describe what's going on with standing waves, left-moving waves and right-moving waves in this heterotic string. I'm going to use this in one more way and tell you a secret. The standing-wave mode, which we have used in the middle, corresponds to one of the strands on the string that describes space. So you need nine of these standing-wave modes to describe the space of the heterotic string, and then you need one more to describe the time. Now it's a standing wave mode, so we can watch it vibrate back and forth. But now we can do something different.

If I took the two pictures on the side—and this kind of looks like Mickey Mouse—so let's imagine, we blew up the two circles on the side and superimposed them on the middle—then all that would be left is, in fact, the middle picture. You'd see a standing wave vibrating both up and down, and nothing else. Can we reverse that process? Can we take a pure standing wave mode and pull it apart into its left and right movers? Here's another way to look at a standing wave that we've used, namely, to take a standing wave of the heterotic string. Remember, when I show you these pictures, I'm actually thinking about equations. So here's a standing wave mode. There is a process known as *Fourier analysis*, which is very well known in mathematics. It goes back to the 18th century, to the French mathematician Fourier. If you apply this mathematical tool to the standing wave, what you can find is that it is exactly equal to a mathematical expression for a left mover, and added to the mathematical expression for a right mover.

So left- and right-moving waves, when added together, gives one standing waves. Remember, in the heterotic string, when we say standing wave, we also mean something that's neither a left mover nor a right mover, but those were the dimensions. Therefore, it means that in string theory, if you think you're looking at a dimension, you have to be very careful, because anything that you think is associated with a standing wave actually can be pulled apart mathematically into left movers and right movers. Why is that important? Well, it turns out that the angular variables, the things that are associated with the change in the photon, are always either left movers or right movers, but they're never standing waves. If you have this understanding of string theory, then it has a kind of flexibility about it, namely, what you think is a dimension can be pulled apart and traded. You go from a theory with, say ten dimensions, to a theory that has less. But in the process, you have to introduce more charges because effectively, what you do is you trade off directions for charge.

This idea is actually a very potent one because it means that, in studying the mathematics of the heterotic string, there are two alternatives, and it's not just the heterotic string. All string theory has this aspect, all closed strings. There are two alternatives to interpret the mathematics that one is looking at. One that is popular is to look at the mathematics and say we are seeing the presence of extra dimensions. By Mr. Fourier's mathematical trick, an alternative version is available, where you can say no; part of what you think are extra dimensions are actually these angular type of variables that are associated with the change of force-carrying particles. So the notion of dimension in string theory itself morphs to become something that is much more flexible. String theory has this really strange aspect of it, that what you think is a dimension may actually be two kinds of charges, left charge and right charge. In the heterotic string, we learned that all right charges go into this number 496, and we were in nine spatial and one temporal dimension.

Now let's imagine an experiment. Let's take one of those nine mathematical objects that we think of as describing the space of the heterotic string, and let's pull it apart as I've described. What do you get then? In fact, let's not stop with just pulling apart one of them; let's pull apart precisely six of them. You go from an object that exists in ten dimensions to an object that exists in four. Well, gee, that looks a whole lot like our world because in our world, we have three directions, front/back, left/right, up/down and time. So the mathematics of the heterotic string turns out to allow one to go down to four dimensions. But in the process, you have to introduce more than 496 charges. Now you might also be worried about the corresponding left movers, because remember I said every time you take a dimension, you pull it apart to get a right piece, which is a charge, but there's a left piece that's left over. What's this left over piece all about?

Well, shortly after the period that I was working on string theory and we found, in collaboration with my friend Warren Siegel, this third score for the heterotic string, we actually pursued this idea of pulling the string apart. When you do that, what you find is another formulation of the heterotic string that has more charge than the original version, and it has something else the original version does not have. When we went through the denizens, I pointed out that nature is very prolific in an interesting way. We have the electron. We have a copy of the electron that's 200 times as heavy; it's called the muon. Then we have a third copy of the electron, which is 1,700 times as heavy; it's called the tau particle. When one writes a string theory in four dimensions, what you find is that the particles in the theory come in copies. In fact, a string theory in four dimensions looks more like our world than a string theory written in higher dimensions.

In some sense, you trade space for charge and copying, but this is a very natural explanation for why copies exist. Why is it that the electron has a copy? Well, without string theory, we have no way to answer this question. With string theory, what we can see that it corresponds to is that, in some sense, it's an extra what we used to think of as direction that has been pulled apart into a charged piece and a piece that is responsible for what we call family member in the Standard Model. So string theories are more complicated than you thought. They're more complicated than most physicists thought, actually. This idea that you can pull them apart, we've described in terms of words, but let me give you a different picture. In fact, most of the people who work in string theory and most of the discussions that you may have heard tell you about this process called *compactification*.

How do I understand compactification? Well, suppose other directions really are out there in some sense. If they're there, let's imagine—we've already imagined that they're small—but let's imagine that they're incredibly warped. In fact, in most discussions of string theory, we find discussions of these warped extra pieces that are called Calabi-Yau Manifolds, at least in their most popular guise. It turns out that physicists are prolific in the ways that they can warp things too, so there are other ways of warping besides the Calabi-Yau Manifold. What's the difference between taking an extra dimension and warping it, and starting with these extra angles? The difference is that you actually have to do something to warp an extra dimension. If string theory really allows extra dimensions, then they ought to explain why you have to warp them. That's something that we have never seen a consistent and complete explanation given in the context of string theory.

On the other hand, if we start from the point of view that string theories ought to describe our world as we see it, then by definition almost, we should allow for string theory to give us the copying processes that we see in the Standard Model as well and allow the string theory to directly let us see the angular variables that are not associated with actual motion, but are associated with the changes in objects such as the photon. If you take this view, then string theory comes to live comfortably in four dimensions. In the four-dimensional point of view, it's not a question about warping because the theory has no options. Whereas if you believe in extra dimensions, the hidden question is why did the theory warp itself? This is a question that we don't have an answer for. Now the actual settling of these questions is going to probably take decades and maybe centuries. Is string theory an actual four-dimensional construct, or is it a higher dimensional construct where you have compactification?

The point of view of four dimensions, I should also make some comments about, additionally. This is not just my work. In fact, in the middle to late 1980s, there were three groups of physicists who seriously studied this question of four-dimensional strings. There was one group led by a physicist whose last name is Antoniadis. He was joined by Bachas, Kounnas, and Windey. They extensively studied the idea that the heterotic string is an actual four-dimensional construct, and they used a particular mathematical approach that is called free fermions. That's a way of saying that we'll work with—remember that we talked about the two descriptions of the heterotic string, that it had a bosonic description and a fermionic description—well, they essentially used all the fermionic description.

A separate group of physicists composed of Dieter Lust, Stefan Theisen and G. Zoupanos used a technique called the *covariant lattice*. The covariant lattice—that's like using the old bosonic formulation totally for the heterotic string that we discussed in the last lecture. Therefore, at least three groups of physicists have concluded that string theory really is—or rather can be—interpreted as a four-dimensional construct. The drawback in this is that all of these objects, as four-dimensional theories, are much more complicated to write than their higher dimensional analogs. Now let me return to the discussion of warping, because I told you that there was more than Calabi-Yau. In most recent years, physicists have been looking at things called G2 compactifications, spin bundle compactifications, and they've added an additional concept called fluxes. These are all different mathematical tools for imagining starting with higher dimensions, rending and bending the space in some way, and getting something that winds up with a four-dimensional description.

There's an interesting question about this. You see, we live in a particular universe. If you start with a string theory in some higher universe, and you want to come down to four dimensions, if you believe in higher dimensions, you have to explain the warping. In particular, you have to pick the warping that agrees with our universe. Well, every few years, a physicist comes along, or a group comes along, with a new way of compactifying. We have no guarantees yet that, if you believe in a higher dimensional point of view, that we have figured out the right way to bend the extra dimensions to produce our world. On the other hand, if you start with the four-dimensional view of string theory, start with the theory where the parts of our world—that is, the Standard MMMModel, which is carriers—are present in the string, and we know how to do that. Then it turns out that the string requires you to add many more parts that we don't see in the Standard Model. That's something we don't quite understand the process for doing. That's, in fact, one of the unsolved problems with string theory. It involves something called modular invariance, a very complicated process.

If one day, however, we discover how to do that, we'll be able to write many copies of the string that contain our world and other stuff. Gee, that other stuff might worry you. In Lecture One, we talked about stuff called dark matter, things that are not like the electron and the proton. Well, in four-dimensional strings, you copiously produce dark matter. In fact, as far as string theory is concerned, it's a good thing that dark matter is out there, because that's the price that we pay for bringing a string down to four dimensions. So are the extra dimensions quark like or string like? We don't know. I want to end this presentation by showing you the difference between the mathematical description of the higher dimensional strings and the four-dimensional ones. You'll see for yourself why so many physicists have chosen to believe that we live in a universe of higher dimensions, but to which you have to bend the extra directions by this process called compactification. I've shown you part of these scores before.

Remember, for the heterotic string as first found by the Princeton String Quartet, their papers contained essentially two scores. I've shown you these two scores. Let me quickly put them up again. That's their first score. That's their second score. Then there's a third score that I personally was involved in with my colleague, Warren Siegel. All of those mathematical expressions I've shown you describe theories in ten dimensions. This last one is everything that you saw in the last transparency, plus this extra stuff. It's more complicated. It's more difficult to write. In many people's minds, the ones that are simplest to write are more elegant, but nature does not necessarily follow our rule for elegance. In fact, Einstein once said, "In the search for truth, leave elegance to the tailor."

Lecture Seventeen
The Fundamental Forces Strung Out

Scope:

In a much earlier lecture, we discussed a dilemma encountered by Stephen Hawking with regard to black-hole radiation. In this lecture, we will consider how superstring theory apparently provides a rigorous mathematical proof to support Dr. Hawking's intuition. In the process, we will note that experimental support for string theory may come, not by looking at the smallest structures in the universe, but rather, by looking at the largest structure, the universe itself.

Outline

I. A prediction of general relativity (the theory describing how gravity works in our universe) was that objects might exist that are so massive that nothing can get away from their gravitational pull. These objects are black holes.

  A. A black hole is not just a singularity but also has about it a sphere. The special property of the sphere is that light can orbit on it, just as satellites orbit around Earth. The sphere is called the *event horizon*.

  B. General relativity also predicts that light can be bent by gravitational forces. This was shown in an earlier animation where starlight was distorted in the night sky as a black hole passed across a field of view.

  C. The gravitational pull of a black hole is so strong that nothing can escape from inside the event horizon. In the late 1960s, Bekenstein took the first step in showing that black holes ought to radiate heat. In a brilliant insight, Hawking asserted that the usual arguments about black holes do not account for laws that apply in the world of the quantum.

  1. Our world is such that if two balls are thrown at each other, they will collide and bounce off one another. But two flashlight beams pass through each other, as if one doesn't sense that the other is there. These are classical expectations.

  2. Using the laws of the quantum world, light and everything else behave differently.

3. To understand the quantum behavior of light, think of a Feynman diagram showing a box with two wavy lines, representing light, attached to the bottom vertices. According to $E = mc^2$, energy can be traded for mass; thus, the sides of the box represent the motion of a particle and an antiparticle (an electron and anti-electron). At the top vertices of the box, the photons stream away. This diagram describes a process called *Delbruck's scattering*, predicted in 1933. According to $E = mc^2$, energy can be traded for mass; thus, the photon can be replaced by a particle/antiparticle pair. The sides of the box represent the motion of this particle and antiparticle. At the top of the box, the particle/antiparticle pairs have recombined to form a photon, and the photons stream away.
4. In 1953, an experiment was performed that showed that light can scatter itself. Remember, the source of the scattering is quantum mechanical processes. This example shows us that to understand our world completely, it is sometimes necessary to take into account quantum processes.

D. In previous lectures, we discussed the idea that the force of gravity, as described by Einstein's equations, does not take into account the workings of the quantum world. Einstein's general relativity, combined with the laws of the quantum world, leads to infinite gravity, which would be a disaster. The definitive demonstration of this was given by a group of physicists in 1974.

II. The problem that Stephen Hawking encountered was that there was no mathematical way to support his intuition that hot black holes (like all other such objects in the universe) must radiate heat.

A. It is useful to review this crisis in physics in a slightly different way.
1. An animation shows why two electrons repel each other. The two balls of light represent electrons, and the wavy line between them represents the exchange of a message carrier (in this case, the photon), telling them to be repelled.

2. In the quantum world, there are more complicated pictures. Again, the two spots of light represent electrons, but one of them emits a photon, which it then later reabsorbs; it also emits a second photon that is responsible for the repulsion.
3. When Feynman Rules are applied to this picture, the mathematics that comes out implies that the force of repulsion between two electrons is not exactly what is taught in high school science.

B. This process can also be applied to the force of gravity. Using the same picture, think of one little spot as the Earth and one as the Moon. In this case, the wavy line used to represent the photon now represents the graviton, the message carrier of the gravitational force.
1. The graviton spins at 4 times the rate of an electron.
2. Applying Feynman Rules to the graviton, the result is infinity. There must be a way to change the mathematics.
3. For now, we have found only one other mathematical way to support Hawking's intuition about the ability of black holes to give off heat. Of course, this other model is string theory. The secret to supporting Hawking's intuition occurred with the photon that was emitted and later absorbed. If the emitted photon traveled outside of the black hole and the electron did not, the photon could be free to carry off the heat!

III. How is string theory different from previous constructions, and why does it seem to work?

A. An animation of two strings moving in space-time shows that they might join together to form a single string that continues moving as a unit. It is sometimes useful to consider the image of a y-shaped figure to show the same process. The two circles to the left of the open end of the y-shaped figure represent the two strings. The two strings move together until they touch each other, then join together and move off as a single string, represented by a single circle on the right side of the y-shaped figure. This y-shaped figure is called the *pants diagram*.

B. In the next animation, a single closed string divides itself into two strings. Again, it can be helpful to consider a still image, with the open end of the y pointing in the opposite direction. A string is indicated with the circle on the left side of the y-shaped figure; it then separates into two strings, represented by two circles on the right of the figure.

C. A third animation can combine these processes. Two closed loops join together to become a single closed loop for a while; this loop then divides itself and becomes two loops again.

D. Because strings are smaller than molecules of air, they don't produce sound when they vibrate; when strings vibrate, the only way for humans to detect them is as the particles introduced as the denizens of the quantum world. As the two strings come together in the animation, it is possible to consider a single vibrational mode of the string.

E. This animation shows the path of vibrational modes as a v-shaped dotted line to the left and right of the wavy line. In the middle, when the string is a single unit, there is a wavy line, because when the strings join together, the modes of vibration change. If studying properties of the electromagnetic force, the carrier in the middle must be a photon. But if considering the force of gravity, the carrier in the middle must be a graviton. It is possible to look at this process as a still picture with the two y's connected in the middle.

F. By applying Feynman Rules, calculations of mathematical expressions from these pictures can be obtained. Further, for every diagram possible for a string, a simple "pinching" of the string diagram yields the same process for particles. Remarkably, every possible way to "pinch" the string diagram leads to a well-defined mathematical expression for a force. It is not separate objects that transmit the forces, then; it is always the string itself. The pinching applies to all possible pictures and allows the "wanderer," discussed above, to carry away the heat of the black hole.

G. As early as 1980, the physicist Daniel Friedan used this idea to calculate from string theory how the forces work. For physicists, this is a calculation called the *effective action*.

1. The ordinary action gives us the effect of all the classical pictures—that is, our first diagram with no complications in the middle.
2. As more complicated interior structures are added to the diagram, they must produce more complicated mathematical expressions. The effective action is what keeps track of all of these pictures. By having all the pictures built into it, a complete description of the force emerges.
3. Friedan showed that Einstein's equations can be derived from the equations of string theory. Further, general relativity is not required to carry out this derivation.

IV. This idea of calculating quantum corrections can be applied to the problem of black-hole radiation.

A. In 1996, two physicists at Harvard, Andrew Strominger and Cumrun Vafa, applied the picture-making technique of superstring theory to try to understand how black holes can produce heat. By using rigorous mathematical rules that follow from the structure of string theory, they were able to produce exactly the result that Hawking had found through intuition.

B. The lesson here is that string theory is the only piece of mathematics known that supports the idea that hot black holes can radiate heat.

V. Many have claimed the existence of rivals to string theory. One of these rivals is called *loop quantum gravity*.

A. Loop quantum gravity starts from a different viewpoint. The loops in this theory are not like the filaments of string theory but can be thought of as plaquettes, flat plates that are used to measure how space is bent.

B. Several years ago, a group of physicists used the formulas of loop quantum gravity to try to calculate Hawking-Bekenstein radiation. The initial results indicated that loop quantum gravity was capable of making this calculation. Within the last year or so, however, the calculations of loop quantum gravity have been found not to agree with Hawking-Bekenstein radiation.

VI. What good is our quantum theory of gravity?
   A. This question cannot be answered yet. To take advantage of this theory in the realm of technology may require advances in technology of several hundred years.
   B. What else does the quantum theory of gravity mean for our universe?
      1. In an early lecture, the cosmic microwave background was introduced. It is seen everywhere in the heavens and represents our universe when it was about 300,000 years old. It looks similar to static on an old TV set.
      2. The average size of the "spots" in this static can be measured to get information that may be useful for studying string theory. Understanding the fine structure in the cosmic microwave background depends on the equations that describe gravity.
      3. As seen in this lecture, the equations of string theory describe gravity in a way that is consistent with the laws of quantum mechanics. Ordinary general relativity does not do so. Thus, there must be attributes in the distribution of the spots of the cosmic microwave background that potentially encode this different description of gravity, if the physics of black holes are a hint.
      4. It's possible, then, that evidence of strings will not be found in the world's most powerful accelerators. It may be necessary to look up at the heavens and try to read the signs of string theory in the structure of the cosmic microwave background.

Readings:

Greene, *The Elegant Universe* and *The Fabric of the Cosmos: Space, Time and the Texture of Reality*.

Kaku, *Hyperspace: A Scientific Odyssey through Parallel Universes, Time Warps, and the 10th Dimension*.

Hawking, *The Universe in a Nutshell* and *A Brief History of Time*.

Smolin, *Three Roads to Quantum Gravity*.

Questions to Consider:

1. Describe the action of a photon carrying a message between two electrons in classical physics.
2. What's different about this picture in the quantum world?

Lecture Seventeen—Transcript
The Fundamental Forces Strung Out

Many lectures ago, we talked about a dilemma that was encountered by Dr. Stephen Hawking. Let me remind you a little bit about what was going on. We had learned that one of the predictions of Einstein's Theory of General Relativity, the one that describes the way that gravity works in our universe—that there might exist objects that are so strong that nothing can get away from their gravitational pull and these objects are called black holes. When we say nothing, we mean absolutely nothing, including light. But a black hole is an interesting object. It's not just a singularity, which is a single point, but it has this kind of sphere around it. The special property of the sphere is that, if you take light, you can orbit it any place on the sphere, just as we orbit satellites about the Earth. This sphere is called the *event horizon*. Anything outside of the event horizon doesn't know there's a black hole around, so physics goes on pretty much as normal.

Another prediction of Einstein's Theory of General Relativity is that light can be bent by gravitational forces. I have a transparency that you've seen before running here. If you were to go out on a starry night, and a black hole passed between you and the distant stars, you would see something like what appears on our image. You would notice that the stars would seem to move around and change their position, and that's actually the light being bent from the black hole as it moves across your field of vision. The other thing that's really interesting about this animation is that black holes classically are so strong that nothing can get outside of the event horizon, but in the late 1960s, a physicist by the name of Bekenstein figured out that that couldn't quite be true. Namely, he figured out that black holes look like they are warm objects in a sense. We all know what happens with a warm object; it gives off heat.

So how can heat get out of a black hole? That was the question. Stephen Hawking, in a brilliant piece of insight, said that what was missing there was that the usual arguments did not account for the laws that apply in the world of the quantum. The paper in which Stephen did this is a marvelous example of how human creativity and imagination are actually more powerful than mathematics, because his paper doesn't actually have a mathematical support for this idea. That was the dilemma, and that was the dilemma we described as Dr. Hawking's dilemma. Dr. Hawking's intuition is actually very interesting because it says you have to take into account what goes on in the world of the quantum, the world of the small.

Our world is such that, for example, if you take two balls and throw them at each other, they bounce off with no problem whatsoever. On the other hand, if I take two flashlight beams and aim one at the other, they pass through each other, and seemingly as if one doesn't know the other is there. These are classical expectations. On the other hand, even if you replace the balls by things such as protons or, more particularly, neutrons, which have no electrical charge, neutrons would also bounce off of each other. The reason then is because they each, instead of carrying electrical charge, carry another kind of charge that we call the weak nuclear charge.

If we apply the laws of the quantum world, light behaves differently. In the quantum world, as we have seen by these pictures called Feynman diagrams, light can be thought of as a wavy line. If you have two light beams, you can think of two wavy lines that are approaching each other. We can draw a box in the middle of them and attach the wavy lines to the bottom vertex of the box. That represents the lines coming in. What is this box that we've drawn, this square? Well, according to $E=mc^2$, energy can be traded for mass. So these sides of the square represent the motion of a particle and an anti-particle. Let's say, to be specific, an electron and an anti-electron. They move along, and on the sides of the box, and then we have the photons, again streaming away at the upper sides.

This diagram is known as *Delbrück scattering*. In 1933, a physicist by the name of Max Delbrück predicted this process. In fact, he predicted it even before Mr. Feynman's pictures had been invented. In 1953, an experiment was performed, which found that light scatters itself. Remember, the source of the scattering is because of quantum mechanical processes. This is an example that shows us that our universe and the world is such that, in order to understand phenomena, sometimes we have to take into account quantum mechanics and quantum processes. The pictures I've been using to talk about the quantum world and how the forces work were actually invented by the physicist Richard Feynman. In fact, they occurred in his work that led to a Nobel Prize, but the prize was not given for the diagrams themselves. It was for actually being able to understand how to make predictions based on the laws of the quantum.

He shared this prize with two other physicists, Julian Schwinger and a Japanese physicist by the name of Tomonaga. Interestingly enough, Tomonaga had actually done his work prior to the work by the other two, but it had not been known in the West; it had only been known in Japan. In fact, Tomonaga's work was done before World War II. Of course, we know what the intervening period was like, where there was little communication between the scientific establishments of these nations. Feynman's rules and Feynman's pictures are a marvelous way to talk about mathematics, and that's why I've emphasized previously that when we use these pictures, we're really doing mathematics in a neat and cool way.

We've also, in previous lectures, talked about the idea that the force of gravity, as described by Einstein's equations, does not take into account the workings of the quantum world. Therefore, if we want to include the quantum world, we must go back to the same sorts of pictures that we've used before. When that happens, disaster occurs. You find out that the force of gravity can be infinite. Of course, that's nonsense. If the force of gravity were infinite, we would all be crushed together in a single super black hole, and that clearly doesn't happen. The actual explicit proof that the laws of the quantum, when combined with Einstein's General Relativity, leads to a disaster was worked out in 1974, first by two physicists, t'Hooft and Veltman, and then all separately in a collaboration that included Deser, Tsao and van Nieuwenhuizen. These direct calculations concerned what the pictures tell us, that when you try to do the picture making for gravity, you find out the force of gravity is infinite.

This was the problem that Stephen Hawking encountered. There was no mathematical way to support his intuition that black holes, since they're warm, have to give off heat, so there was a problem to be solved. He wrote this very famous paper that, if you look at the mathematics in the paper, the mathematics doesn't support the ideas. That's not how physics works. We're not allowed to simply say, "My intuition is such that I predict this." Most physicists doing that will find their colleagues are highly skeptical. In the case of the black hole, even though there was room for skepticism, there was no proof that Stephen Hawking had it wrong. Just as important, there was no proof that he had it correct either, so there was a crisis. This crisis, perhaps more than any other piece of physics, is responsible for many people starting to think about something such as string theory.

Let's review this catastrophe in a slightly different way. I have now running a diagram that, in some sense, explains why two electrons repel each other. The two balls of light represent electrons. The wavy line in the middle is the exchange of a message-carrying particle saying you ought to be repelled. Since we're talking about the electromagnetic force here, that message carrier is the photon. On the other hand, in the quantum world, we can include a more complicated picture, and here we have one. Again, the two dots of light represent electrons. You'll notice that on the left-hand side of this diagram, the electron emits a little photon, which it later reabsorbs, traveling along the path of the first wiggly line. Then it's an emission of a second photon that's responsible for the repulsion.

Mr. Feynman's rules tell us how to turn this expression into mathematics. When you apply Mr. Feynman's rules to this picture, you find that the force of repulsion between two electrons is not what you are taught when you enter a high school class in science; it's something quite different. Of course, there are many, many other pictures that may apply, and each picture makes a slightly different prediction about the force of repulsion between two electrons, so that's the general process. This process may be applied to the force of gravity. We may take that exact same picture that we just discussed. Instead of thinking about the little balls of light as being two electrons, you may think about one being the Earth and the other being the Moon.

In this case, we are interested in the gravitational attraction. The wavy line in the picture would no longer be a photon. It would be the message carrier for the gravitational force, the object we call the graviton. You remember the graviton. We met it when we went to the world of the quantum. It's an object that spins at four times the rate of the electron. If we try to use Mr. Feynman's rules for the graviton, that's what leads to infinity. So there's something wrong with the mathematics here. We have to find a way to change it. In fact, the statement is, as of right now that we have only found one other piece of mathematics that supports Dr. Hawking's intuition about the ability of black holes to give off heat. This other piece of mathematics is string theory. We want to see how string theory is different from previous constructions and why they seem to work.

We can begin again with an image. These two loops that you see here represent two strings, but you'll notice I allow them to move and join together to form a single string. It's complicated to see in this animation, so let me put a still up on the screen to show the same thing. We see the two circles at the ends on the left over here; those represent the two strings. The two strings move together, and at this point, they join, touching each other, and then move off as a single string or a single unit. That's what the picture showed. If that process is allowed to occur for strings, what if we started with a single closed string and allowed it to pinch itself off into two pieces? That's what we're watching in this animation. It will repeat itself, so let's watch it again. There's one string. It pinches down and becomes two, and then the two separate pieces move off. Again, that's kind of busy, so we can draw a simpler picture, which contains the same information. Here it is. We have a single string on the far left-hand side of the picture. It's moving along, and then it splits into two circles, which represent our strings. Then those move off on the right-hand side of the picture.

If those two things can happen, let's join them together to see the result of that. This is an animation that we see running here, that carries this process out. We begin with two closed loops. They move together, and then they join and touch and become a single closed loop, which then, after a while, splits apart. We can watch this process. I have it in repeat mode, so we can see it over and over again. There it is. Now we can ask a very interesting question. Remember, we've talked about the fact that the kinds of strings that are the subject of this lecture series are so small that, when they vibrate, they don't produce sound because they're smaller than the molecules of air, which cause sound to be accessible and occur in nature.

When these objects vibrate, the only way that we would detect them is as particles that we met in our journey to the realm of the quantum. So we can ask a question, as the two closed strings are coming together—let's watch a single vibrational mode—why? It's because a single mode of vibration of the string looks, to our level, like a particle. We'll watch a single vibrational mode and draw its path. If you look in the background of this picture, that's actually what I've done. You'll see that on the left, there's a line; that's the path of a single vibrational mode. It looks like it's V-shaped. First it comes in, and then it moves away. The same is done on the right-hand side. So those are pictures of the path of a single mode. The middle, when the string is a single unit, you'll see there's a wavy line. The reason is because, in order to join together to become a single string, the modes of vibrations of the strings have to change. If you look at what's happening with a single mode, you no longer have the same object in the interior of this picture that you have in the exterior.

If we're talking about the electromagnetic force, the carrier in the middle must be a photon. On the other hand, if we're talking about the force of gravity, the carrier in the middle must be a graviton. Here's a simpler version of that process. We see two strings on the bottom coming together, forming a single closed string then moving off again. The remarkable thing about these pictures is that, applying Mr. Feynman's rules, we can calculate mathematical expressions from these. So we can find out from these pictures of strings, what is the force between the Earth and the Moon? If we apply the classical picture, which is only the simplest picture that we saw. If you look in the upper right-hand part of this diagram, you'll see we've reproduced our picture of the two closed strings moving together, moving as a single unit, and then splitting again to two. Compare that to path of a single note, and we find out that that's the classical picture that, for the particle, leads to the 1 over r-squared law that we're familiar with.

Under the dashed yellow line, you'll see that there're some more pictures. On the particle side, we've seen the exchange where there's an extra photon on one leg, on the left-hand side of the diagram. We can still see that here. When we go to consider the string, the way that we reproduce that result is right on the left side of the string diagram. You should imagine that this diagram is flexible; it's like a soap bubble. I can pinch it at a certain point, and there is, in fact, a pinched hole on the first string diagram. That represents the effect of the photon exchange. In the second string diagram, we pinched the hole in a different place. The idea is that for every diagram that we can draw for a particle, we simply pinch a different kind of hole in the string diagram. The most remarkable part of the process is that, for the string, when we look at the mathematics, every way that we can pinch this figure off leads to an expression for the force that's well defined. Why is that important? Well, let's continue our story about the black hole.

In order to transmit the forces, what happens is that, well, it's not separate objects that are doing it. It's the string itself. It's always the string either vibrating in a different way or moving in a different way, but it's not something different or new coming into play. It's just the string. The process that we've seen in this animation is something that represents a series of mathematical calculations. As early as 1980, a physicist by the name of Friedan used this kind of idea to calculate from string theory how the forces work. For physicists, this is a calculation called the effective action. What does that actually mean? Well, the ordinary action basically gives us the effect of all the classical pictures. That's our first diagram with no sort of complications in the middle. Then as we begin to introduce more and more complicated interior structures to the diagram, we have to produce more and more complicated mathematical expressions. So the effective action is the thing that keeps track of all our pictures. By having all the pictures built into it, it's a complete description of the force.

The remarkable thing about Friedan's work is that he showed that if you take the equations of string theory and look at their pictures and calculate the mathematics that's associated with those pictures, remarkably enough, you may derive Einstein's equations. What's really strange about that is the following. You don't have to know general relativity in order to carry out this derivation. In fact, what this work shows is that if Einstein had never lived, or if Einstein had only given us the Theory of Special Relativity, but not the Theory of General Relativity, to explain gravity, string theory, if it had come into existence, is completely capable of providing a derivation that was given to us by Einstein's genius in 1916. Maybe 40 or 50 years later, we might have bumped into it by studying string theory. Notice that string theory follows from following the route of the quantum.

In the case of Einstein, remember he didn't like the world of the quantum. This was something that, throughout most of his life as a physicist, he opposed as a concrete and fundamental description of nature. He thought, and in fact we've often heard this quote made, that God doesn't play dice with the universe. Apparently, God does, because quantum theory is the best-tested piece of science our species possesses. We've checked the accuracy of the predictions of this mathematics to greater than ten decimal places in a certain experiment. The laws of the quantum world govern our universe.

This idea of calculating the quantum corrections means that you may apply it to the problem of the black hole radiation. This was actually a very recent realization. In fact, it only occurred in 1996 when two physicists at Harvard University, Strominger and Vafa, applied the picture making technique of superstring theory to trying to understand how it is that black holes can produce heat.

The thing that's really interesting about this story is that Hawking had to make a guess, but Strominger and Vafa simply made some calculations. They took a very special string, something that's called a type IIB string—we're going to meet this string in a later lecture and understand what it is—it's a special string. They took a limit of it where the mathematics was in control, and then what they did was to count the ways that the entropy—remember when we talked about black holes, we did those very funny games of calculating probability. When you have probabilities, you can use those to calculate what we physicists call entropy. Strominger and Vafa actually made this calculation.

What they found, most remarkably, is that by simply using rigorous mathematical rules that follow from the structure of string theory, they were able to produce exactly the result that Stephen Hawking had found. Remember, Stephen didn't find them. What Stephen actually did was to say, "My intuition tells me. So we have an example where the mathematics of string theory, which occurs almost 30 years later, catches up with the intuition of a great physicist. That's, in fact, the mark of a great physicist—the ability to move beyond the equations that we use in order to describe the way the universe works. So the lesson here is that string theory turns out to be the only piece of mathematics that we know that can support the idea that black holes are not totally black, but are capable of giving off heat.

Many of you, perhaps, have heard of rivals to string theory. One of these rivals is known as loop quantum gravity. Loop quantum gravity starts from a very different viewpoint. Remember, the idea in string theory is that, at the very most basic level of our universe, there are these tiny filaments. You hear the word loop in loop quantum gravity, so it sounds like it's the same, but it's, in fact, very different. The loops of loop quantum gravity, in some sense, can be thought of not as tiny loops that are the fundamental entities in the universe, but instead are objects that are varying plaquettes. A plaquette is a flat plate that you can use to measure the way space is bent. Remember, Einstein tells us that gravity is really the curvature of space.

For many years, people have thought that perhaps loop quantum gravity would be a rival to string theory. In particular, several years ago, a group of physicists used the formulism of loop quantum gravity in order to try to calculate Hawking Radiation. That's the fact that black holes are slightly warm. More properly, we should refer to this as Hawking-Bekenstein Radiation. The initial results were that loop quantum gravity was capable of doing this. In fact, most recently, within the last year or so, the calculations upon which this was based were reexamined by experts in that area. To their dismay, they found that the calculations of loop quantum gravity do not agree with the idea that Hawking has given, that black holes can radiate. As of right now, this means that one of the main competitors to string theory looks less attractive to scientists than it did three or four years ago.

The basic lesson here is that we now have a quantum theory of gravity. What's a quantum theory of gravity actually going to be good for? Well, I can't answer that question because, you see, in order to take advantage of loop quantum gravity at the level of technology, we're going to have to advance, probably, several hundred years before that will occur. I remind you that this is like what happened with the electromagnetic force. It was in the 1870s that Maxwell wrote his equations to describe it. It was only in the last 10 or 15 years that we've had nice devices such as cellular telephones. That's on the order of 120 to 140 years. So the technology that flows from this will likely not be something that we can understand or realize to make our lives better or our grandchildren's lives better for maybe several generations.

What else does this mean, though, about our universe, that we have this quantum theory of gravity? In one of my early lectures, I talked about the fact that there's a fossil that's written in the heavens. It's called the cosmic microwave background. The cosmic microwave background is everywhere that we look, and it represents our universe when it was about 300,000 years old. The cosmic microwave background looks like static if you look at pictures. If you've seen static on your television screens—of course, this assumes that your TVs are not digital—but if you've seen static on the old analog television screens, that's what the signals from the cosmic microwave background look like. They're just distributed in what looks to be a random fashion, but the interesting thing about this pattern is that you can measure the average size of spots and gain information.

This process of measuring the average size of the spots in the cosmic microwave background is something that perhaps we can use to study string theory. Why? Well, it turns out that this fine structure in the cosmic microwave background depends on the equations that describe gravity. We have seen in this lecture that the equations of string theory describe gravity consistent with the laws of quantum mechanics. Ordinary general relativity does not do that. So in some sense, there must be attributes in the distribution of the spots of the cosmic microwave background that potentially can encode this different description of gravity. So it's completely possible that instead of finding evidence for things such as strings by going to the world's most powerful accelerators such as we've discussed previously, the Large Hadron Collider, it may well be that instead we should be looking up toward the heavens and trying to read the signs of string theory by looking at the structure of the cosmic microwave background.

The fact that maybe the answer is in the heavens as opposed to looking in powerful accelerators is an example of a marvelous coming together in our understanding of physics. In order to understand things that are in the heavens, over the last decade we have come to understand, it is absolutely necessary that we understand the world of the quantum. We may now take this one step further, perhaps. If we find the signals that describe the influence of superstrings on the cosmic microwave background, then we may say with some certainty that the signs of string theory, the thing that many people dispute as being unproveable and untestable, that those signs may well come down to us by looking up at the heavens. I believe that would be a marvelous way for the theory to come into the world of physics, to leave the word of mathematical speculation and become an element that we can use to begin to understand this marvelous place in which we live, our home, the universe.

Lecture Eighteen
Do-See-Do and Swing Your Superpartner—Part I

Scope:

Prior to string theory, space and time were separate from the forces and matter that exist in the universe. String theory gives us a different viewpoint, in which space and time are actually part of everything else and vice versa. This lecture will concentrate on the other "stuff" in the universe, namely, the parts that lead to creatures like humans. Mathematical evidence pointing to the existence of *superpartners* for this ordinary matter will be found, and we will also confront the fact that these superpartners have yet to be seen in the laboratory.

Outline

I. Our universe contains objects, the things that make up matter, such as quarks and leptons. The other part of the universe that leads to creatures like humans consists of the forces, which hold the fundamental building blocks of matter in fixed patterns.

  A. A neutron, when separated from the nucleus of an atom, on average, after 1013 seconds, is replaced by a proton, an electron, and a neutrino. The initial neutron was neutral; the proton present afterwards is positively charged; and the electron is negatively charged. This process is called *beta decay*, and the form of energy responsible for beta decay is the weak nuclear force. Other forces include electromagnetism, very familiar to many people, and the strong nuclear force.

  B. All the matter objects we have discussed have the same spin rate as the electron. They can spin clockwise or counterclockwise at a rate of 1/2 h-bar.

  C. The force carriers have spin rates that are very different. The graviton spins at 4 times the rate of the electron. The photon, the intermediate vector bosons, and the gluon all spin at 2 times the rate of the electron. The Higgs particle doesn't spin at all.

  D. Spin indicates which particles obey the laws of chemistry and which don't. Electrons gather around atoms in shell structures that are governed by spin. Since quarks have the same spin as electrons, their shell structure would have to look very much like an atom. In contrast, trying to build an atom with photons, the shell structure would disappear, because anything that spins at 2 or a multiple of 2 times the rate of the electron does not obey the Pauli exclusion principle.

II. Why is it that the forces don't obey the Pauli exclusion principle and can never lead to a shell structure, whereas all the matter fields do obey the Pauli exclusion principle and will always lead to something like a shell structure for atoms? We might think our universe would be more balanced if some of the force carriers obeyed the Pauli exclusion principle and others did not and if matter particles behaved in the same way.

   A. It is useful to look at a graphic indicating the parts of our universe: the quarks (red), the leptons (blue), and the force carriers (green). In an earlier lecture, we introduced the property of supersymmetry to banish the tachyon from the mathematical structure of the string. In supersymmetry, for every bosonic strand of the string, there had to be a corresponding fermionic strand of the string.

   B. Is our world supersymmetric? We can find out simply by counting objects in the standard model, beginning with 18 quarks (6 flavors $\times$ 3 colors) + 6 leptons = 24 fermions. In fact, each fermion counts as 4 objects, accounting for the spin-up and spin-down varieties and for the fact that each fermion must also account for a particle and an antiparticle. Thus, the number of fermionic degrees of freedom in our universe is 24 $\times$ 4, or 96.

      1. Next, it is possible to count the force-carrying particles: 8 gluons + 1 photon + 3 intermediate vector bosons + 1 Higgs. Again, the counting is a little bit more complicated. The photon counts for 2 states, because it can spin to the right or to the left like an electron; thus, the photon has 2 degrees of freedom. This is also true of the gluons. Intermediate vector bosons have 3 degrees of freedom. Adding them all up, we get 29 bosonic degrees of freedom. (For now, we ignore the Higgs boson since no experiment has yet seen it.)

2. Our universe has no hope of being supersymmetric, because 96 does not equal 29 (and including the Higgs particle does not help).
   C. This result presents a dilemma that involves the vacuum.
   1. The vacuum is a state of lowest energy. Accounting for quantum processes, the concept of the vacuum in the world of the quantum is not as simple as might be imagined.
   2. A common concept is that there is nothing in a vacuum. However, in a quantum mechanical universe, particles can spontaneously come into existence if they come as particle/antiparticle pairs. Rather than thinking about the vacuum as a state of nothingness, it must be accepted, from the principles of quantum mechanics, as a sea in which particles pop in and out of existence.
   3. Thus, in quantum theory, the vacuum has a much more sophisticated structure. This raises a question of how to calculate the energy of a vacuum. Using the mathematics of quantum theory, this is a very difficult question…with one exception. It turns out that if the universe is truly supersymmetrical, then the vacuum can be shown to be a state of lowest energy.
III. This presents a dilemma. By counting, the universe is seen to be not supersymmetrical, but the vacuum would necessarily be the state of lowest energy if it were.
   A. The weak nuclear force is the form of nuclear energy that turns a neutron into a proton. When this process was first discovered, available technology was such that measurements of the neutron, the proton, and the electron—but not neutrinos—were possible. Laboratory results showed that the energy after the decay did not equal the energy before. This was also true for momentum. If this result were true, then energy and mass are not the same thing, which contradicts $E = mc^2$.
   B. Wolfgang Pauli asserted the reason energy and momentum appeared not to be conserved in this process is that something was escaping the detectors. This "something" is now known as the neutrino.

C. The lesson here is that sometimes it is possible to imagine a slightly different universe mathematically, but later, experiments must support this different universe.

IV. It is possible to imagine a supersymmetric universe.

A. In a supersymmetric universe, the matter and force carriers known to exist thus far must have supersymmetric partners. In our universe, if we have, for example, the graviton, which has a spin rate of 2, then in a supersymmetric universe, there must also be an object that carries the gravitational force but spins at a rate of 3/2. Such objects have never been seen in the lab.

B. Further, in our world, the electron exists, which spins at a rate of 1/2. In a more balanced world that is supersymmetric, there must be an object that spins at a rate of 0 (the *selectron*). Up and down quarks also spin at a rate of 1/2 in the supersymmetric world; up and down *squarks* spin at a rate of 0.

C. In the graphic there is a balance on both sides. In a supersymmetric universe, for every matter particle and message carrier known, there is a superpartner.

D. Of course, this whole idea is mathematics; it is possible to describe objects never before seen, but if they exist, calculations show they would serve to stabilize the vacuum.

V. Energy is a conserved quantity, and it is expected that there is a state when it is at its lowest value, the vacuum. How do those facts relate to the absence of supersymmetrical particles?

A. To answer this question, we might say that the supposition that the universe is supersymmetrical is incorrect. In string theory, the property of supersymmetry was needed to banish the tachyon.

B. Supersymmetry at the level of the string asserts that the number of strands of the bosonic variety should be equal to the number of strands of the fermionic variety. But the particles in our world are not the actual strands of the strings; the particles are the objects that correspond to the notes produced by the strands. It is conceivable that even though the strands are supersymmetrical, the notes do not come in pairs. A few mathematical string models have this property, but most of the mathematical constructs that come from string theory have both the strands and the notes appearing in pairs. Thus, most physicists think that we ought to be able to find the superpartners.

VI. Why have superpartners not been found?
   A. To see these superpartners, technology that is capable of producing them is required. When this will occur will depend on how massive these objects are.
   B. Typically, the mass of the superpartners is estimated to be 300–1000 times greater than the mass of ordinary matter.
   C. How fast is technology advancing? Consider the plot of the masses of the particles versus the history of detecting particles. The resulting straight-line plot looks like Moore's Law, which states that the power of computers doubles once every two or three years.
   D. This same law can be applied to our ability to detect increasingly massive particles. We progress at a rate of about 3 GeV (giga-electron-volts—a unit of energy) every year. If superpartners are 300 times as heavy as ordinary matter, then we might produce one in a laboratory around the year 2016. But if they are 1000 times heavier than ordinary matter, producing one in the laboratory will take 200–300 years.
   E. The supersymmetric structure also depends on rigorously supporting experimental physics.
      1. In our nation, about 10 years ago, the community of particle physicists had plans to build the Superconducting Super Collider. Ultimately, it was decided that this device was too expensive for the scientific budget of our nation to support.
      2. Europeans have invested in a device about one-third as powerful as the Superconducting Super Collider would have been. This is the Large Hadron Collider (LHC), which is scheduled to be completed in 2007.
      3. The United States has spent approximately $500 million to put detecting devices in the LHC. In this way, Americans will participate in the search for what lies beyond the standard model, and many physicists believe that supersymmetry will be found in the LHC.

F. In 1980, while visiting the University of California at Berkeley, I met Bruno Zumino, one of the physicists who first introduced the idea of supersymmetry in the Western literature. This was also the same year in which the mathematical idea of supersymmetry was applied to the standard model.

  1. Zumino, one of the fathers of supersymmetry, was not as excited about this prospect as might be expected. Sometimes, ideas in physics turn out to be correct, but not in the way we think they are.

  2. In the standard model, there's a special set of equations to describe gluons and intermediate vector bosons. These are the Yang-Mills equations, named after the two physicists who wrote them in 1954. Initially, when these equations were presented, they were created to understand forms of matter related to protons and neutrons. The equations failed miserably at this task, but 20 years later, they worked remarkably well for the weak intermediate vector bosons. About 10 years later, the intermediate vector bosons were found in the laboratory.

  3. It is possible that supersymmetry might follow a similar route. Even though many who have studied supersymmetry find the mathematics to be elegant and convincing, until supersymmetry is observed in the lab, it will not be known if the mathematics describes our world. It is possible this mathematics is being applied incorrectly, but I believe that, if so, these equations will return, maybe 100 years from now, and describe something else of great importance.

Readings:

Gell-Mann, *The Quark and the Jaguar: Adventures in the Simple and the Complex*.

Lederman, *The God Particle*.

Lederman and Hill, *Symmetry and the Beautiful Universe*.

Oerter, *The Theory of Almost Everything: The Standard Model, the Unsung Triumph of Modern Physics*.

Zee, *Fearful Symmetry: The Search for Beauty in Modern Physics*.

Questions to Consider:

1. When we counted the degrees of fermionic and bosonic freedom in our universe, what did this tell us about supersymmetry?
2. Why have we not yet seen any of the superpartners?

Lecture Eighteen—Transcript
Do-See-Do and Swing Your Superpartner—Part I

Our universe, as we have noted on a number of previous occasions, is partitioned in a particular kind of way. First of all, actually we may set aside one aspect of our universe, namely space and time. They play a role on a stage. Prior to string theory, space and time really were separate things from the forces and the matter that exist in the universe. String theory, in fact, gives us a different viewpoint, where space and time is actually part of everything else, and vice versa. We're going to concentrate on the other stuff, namely, the parts in the universe that lead to creatures such as us. There are two things that we may speak on in that regard. There's matter; we've been through a long list of the objects that make up matter. There are the quarks; up, down, charm, strange, top and beautiful. There are the leptons; the electron, the muon, the tau particle and their associated neutrinos. Those are matter.

Then there is the other part that's important for constructing beings such as us, namely, we have to take these fundamental building blocks and hold them in fixed patterns in order to have our parts. The forces are four in number. There is the gravitational force. It's the weakest, but in some ways the most important, because the large-scale structure of the universe is governed by gravity. There is the weak nuclear force. The weak nuclear force—the simplest way to think about it is that if you take a neutron and separate it from the nucleus of an atom, on average, after 1,013 seconds, the neutron is replaced by a proton, an electron and a neutrino. The neutron that you began with was neutral. The proton that you have afterward is positively charged. The electron afterward is negatively charged. So you had to have those two occur; otherwise, you would not conserve charge in this process. Then there's the other relative to the electron, the neutrino, which also comes out of this decay.

This process is called beta decay, and the form of energy that's responsible for beta decay is the weak nuclear force. There is electromagnetism. We're, of course, all familiar with electromagnetism. That's what our high technology utilizes in order to raise standard of living and basically create a way to go about life that was unimaginable, perhaps, in the beginning of the last century. Finally there's the strong nuclear force. We have matter and the forces that act upon them. If we restrict ourselves to only matter—

and I'll call the forces energy; I'll use the words interchangeably—there's something else that's rather obvious. All of the matter objects have the same spin rate as the electron. They can spin counter-clockwise at a rate of one-half h-bar, or they can spin clockwise at the rate of one-half h-bar.

The force carriers, on the other hand, have spin rates that are very different. The graviton spins at four times the rate of the electron. The photon, the intermediate vector bosons and the gluons all spin at twice the rate of the electron. The Higgs Particle, which we haven't said very much about, but it plays an important role that we will come to, actually doesn't spin at all. It's a spinless object. The reason why I point out spin is because that tells us what obeys the laws of chemistry and what doesn't. Electrons gather around atoms in shell structures. This shell structure is governed by spin. If you imagine, for example, gathering quarks in some kind of a structure, it would have to look very much like an atom because the quarks have the same spin. On the other hand, if you imagine looking at—and we can imagine because we can do it with mathematics—you can imagine trying to build and atom with photons. The shell structure disappears because anything that spins at two or a multiple of two times the rate of the electron does not obey the Pauli Exclusion Principle.

There's a problem that we may perhaps think about, and that is why is it that all of the forces do not obey the Pauli Exclusion Principle and may never lead to a shell-like structure; whereas all of the matter fields do obey the Pauli Exclusion Principle and would always lead to something like a shell structure for atoms. After all, would not our universe be more balanced if some of the force carriers obeyed the Pauli Exclusion Principle and others did not, while other of the matter particles would be some obeying the Pauli Exclusion Principle and others not? Wouldn't that, after all, be more balanced? The answer is intuitively yes, so let's go back and look a bit at our universe and its building blocks as we assemble them. Here are the quarks, the leptons—up, down, charm, strange, top and beautiful—and we'll assemble together the force carriers. All the force carriers are green. We have the quarks and the leptons in either red or blue. That's what our universe looks like. This is the modern version of the table of elements for how to build a universe in your basement.

In our discussion of the string, we came across this very peculiar property called supersymmetry, and we needed supersymmetry in order to banish the tachyon from the mathematical structure of the string. In order to have supersymmetry, for every bosonic strand of the string, there had to be a corresponding fermionic strand of the string. If you want an analogy, in order to banish the tachyon, it's like having people coming to a bar. If they come in pairs, the bouncer lets them in, but if they come singly, the bouncer tells them they can't come. That's how supersymmetry works. It says things must come in pairs, where one member of the pair is a boson or a boy, and the other member of the pair is a fermion, a female. So is our world supersymmetric? That's something we can do simply by counting because, if supersymmetry is around, there should be equal numbers. Let's do some counting of the Standard Model.

We have 18 quarks. Remember, they come in six flavors times three colors. We have six leptons, so that yields a total of 24 fermions in our universe. We can't actually use the number 24 because that's not actually correct in the counting. We have to be a little bit more careful. Remember the electron actually comes in two varieties, spin up and spin down. So instead of counting the electron as one object, we should at least include the effects of spin. The other thing about the electron that we must take into account is that it has an anti-particle. In fact, an electron really counts for four different objects—spin up, spin down, particle and anti-particle, and then spin up and spin down for the particle and anti-particle. Those are four objects. So really, we must take this 24 and multiply by four in order to count what we physicists call the fermionic degrees of freedom in our universe. The answer turns out to be 96. We have 96 fermions in the universe.

On the other hand, we can count the force-carrying particles. There are eight gluons, one photon, three intermediate vector bosons and the Higgs Particle. We can count those; however, the counting is a little bit complicated because the photon counts for two states. If you remember our discussion of light, we said it could circle to the right or circle to the left. That's actually two ways that it can count, so we say that the photon has two degrees of freedom. This is also true for the gluons. They have two degrees of freedom. On the other hand, the force carriers for the intermediate vector bosons actually have three degrees of freedom. So when you add up all the ways that objects can spin and their anti-particles for the boson, you find 29 bosonic degrees of freedom in our universe. Well, 96 is not equal to 29, and therefore our universe has no hope of being supersymmetric, as we have seen in the laboratory thus far. It looks as though we don't live in a supersymmetrical universe.

There are reasons why we would want to live in such a place. One of these reasons has to do with energy. You would like to believe, and certainly we scientists believe, that there's a state of lowest energy. In physics, this state is called the vacuum. When you look at the concept of the vacuum in the world of the quantum, it is not as simple as you might imagine. You might think the vacuum means there's nothing there. But in a quantum mechanical universe, particles can spontaneously come into existence if they come as a particle/anti-particle pair. That's what a quantum process is all about. So rather than thinking about the vacuum as a state of nothingness, if you accept the principles of quantum mechanics, you have to think of it instead as more like a writhing sea, where particles are popping in and out of existence. That's what you mean by the vacuum. It's something of a much more sophisticated structure. So the question becomes how do you calculate the energy of the vacuum?

Using the mathematics of quantum theory, this is a very difficult question, with one exception. It turns out that if the universe is truly supersymmetrical, then the vacuum can be shown to be a state of lowest energy, absolutely a state of lowest energy. That's what you really would want for a vacuum. A vacuum should be a state where you can't remove any more energy from the system. There's a very practical reason for thinking that the universe might be supersymmetric, but on the other hand, I just gave you guys a count, and we found the universe was not supersymmetric. There's a dilemma here. How do we resolve this dilemma? It has been now about 100 years or so that physicists have been thinking about the world in terms of fundamental particles. One of the things that has occurred is that, when you see something funny in the laboratory, we have become used to the idea that maybe we can imagine something that's better.

I earlier told you about the weak nuclear force as being that form of nuclear energy that turns a neutron into a proton. Back when this process was first discovered, our technology was such that we could measure the neutron, we could measure the proton, and we could measure the electron, but we couldn't measure neutrinos. That les to a bit of a problem because, if you actually watched this process, you would watch the neutron disappear, and then out would come flying one proton and one electron. You could measure the momentum of each because we could measure their velocity. We knew their masses. You could also measure the energy that they contained because motion contains a kind of energy known as kinetic energy. When you did this in the days of the original experiments, a great crisis arose because if you took the energy after the decay, it didn't add up to the energy before. This is also true for the momentum. If that were true, it means that energy and mass are not the same thing. But that's what Einstein's famous equation, $E = mc^2$, was all about.

The physicist named Pauli entered this debate and said the reason that we think that energy and momentum are not conserved is because something is escaping our detectors. This something, we now know, is the neutrino. The moral of this story is, if you want something to be true, we're allowed mathematically to imagine a slightly different universe. But then we must later come back and ask the question, is this different view of the universe something that can be supported by an experiment. Now let me take you to a supersymmetric universe. In a supersymmetric universe, in addition to the ordinary up, down, charm, strange, top and beautiful quarks, along with the electron, the muon, the tau particle and their associated neutrinos, there must be other objects that are their mirror pairs because symmetries ultimately are always about some kind of a balance.

We have a favorite movie for showing this; it's our ladder. We know that if we have our ladder lying flat, then we can draw two squares that represent the area—one where the ladder is the base, the other where the ladder is the top. So the green ladder here has an area that's always the same as the area of the blue, but we can also take our ladder and put it precisely against the wall, in which case the area of the green then is equal to the area of the brown. So if you want to have symmetry, you have to have some kind of sense of balance. In our supersymmetrical universe, therefore, in addition to the ordinary matter and force carriers that we've seen, we have to allow for the appearance of a new set of objects that have never been seen in the laboratory. There's the graviton, which we know from Einstein's equation of general relativity. If the universe is supersymmetric, there's an object that also carries the gravitational force. It's the message carrier for the gravitational force, but it spins at the rate of three-half. The rate of the graviton is a spin rate of two.

In our world, we have an electron. The electron spins at a rate of one-half. In a more balanced world that's supersymmetric, there must be an object, which we call the *selectron*, which spins at rate zero. In our world, there are up and down quarks, each of which spin at a rate of one-half. In this more balanced supersymmetrical world, there must be up and down squarks whose spin rate is zero. So if you look at this final picture, you can see it is balanced, and that's what we mean by symmetry. You can think of our ladder example where, once again, we saw mirror images of certain squares. This is what a supersymmetrical universe would look like. For every piece of matter, every particle of matter that we know, and every message-carrying particle we know, there would be what we now call a *superpartner*. The superpartner to the photon, which is the message carrier for electromagnetism, would be an object called the *photino*. The superpartner to the electron would be the selectron, and so forth. The superparticle to the Higgs would be called the Higgsino.

Now this is all mathematics, and this mathematics was invented in the late 1960s and early 1970s. But the thing that's interesting about this mathematics is we've never actually seen these objects. We've never seen the things that would, if they're there, stabilize the vacuum, for example. They would also have the role of making our universe a more balanced place. Energy, as I said, is a conserved quantity, and it should be conserved and have a vacuum. So what do we do about the absence of these particles? There are two things. First of all, you can say that this supposition that the universe should be supersymmetrical is, in itself, wrong. Let's go back and review the string, because we know that for the string, in order to banish the tachyon, we absolutely had to require supersymmetry. Remember, that's with regard to the strands. Remember, ultimately I have revealed that the string is not a single object. You should think of it really like a musical instrument. So supersymmetry at the level of the string essentially says that the number of strands of the bosonic variety should be equal to the number of strands of the fermionic variety.

But the particles in our world are not the actual strands of the strings. The particles that we know are the objects that correspond to the notes that the strands produce. It's conceivable that even though the strands are supersymmetrical—namely, come in pairs—maybe it's the case that in our universe, we do not actually see the notes in pairs. There are a few such models, a few such mathematical constructs that have this property. Some of these constructs precisely come from strings. On the other hand, most of the constructions that one can get from string theory not only has the strands of the strings coming in pairs; they also have the notes coming in pairs. The vast majority of the community of physicists who study string theory also are committed to the idea that we ought to—in our world, things that should be accessible to us—ultimately be able to find these superpartners. But we haven't so far. Why not?

Well, there are several possibilities. One is simply a matter of mass and a matter of energy. You see, in order to see these superpartners, we have to have the technology that's capable of producing them. We've never actually achieved that level of technology so far, and it also will depend on how massive these objects are. Why does that play a role? If we make a guess about the mass of these missing objects, the typical numbers are someplace between 300 and 1,000 times heavier than ordinary matter. The other thing that's interesting is how fast our technology is advancing. There's a way to estimate this, which is simply to go back and look at the history of how we detect particles and plot their masses. There is a simple assumption that we can make about making such a plot. Approximate this plot by a straight line, and you get something that looks remarkably like what's known as Moore's Law.

I don't know if you've heard of Moore's Law, but roughly speaking, it says that the power of computers doubles once every two or three years. This has been true in the computer world for over 20 years, which is why computer technology continues to improve. We can apply this same kind of logic to our ability to detect more and more massive particles. If you look at the history, you find out that we progress at a rate of about three GeV—that is giga-electron-volts, which is a unit of energy—on a year's basis. What that means is the following—that if superpartners are on the order of 300 times as heavy as ordinary matter, then perhaps around the year 2016, someone in a laboratory will produce one of these objects. On the other hand, if these objects are 1,000 times as heavy as ordinary matter, then it will take something perhaps like 200, 300 years to actually produce them in a laboratory. So we have to count on nature being very kind to us to reveal the supersymmetric structure of the universe.

The other thing about this supersymmetric structure is that it will also depend on our rigorously supporting experimental physics. In our nation about 10 years ago, the community of particle physicists wanted to build a device called the Superconducting Super Collider. Ultimately, it was decided that this device was too expensive to support on the scientific budget of our nation. This device is not coming back. As I often tell friends, it had a stake driven through its heart, so we don't have to fear the Superconducting Super Collider, or so-called SSC, coming back to cause disturbances with our nation's finances. On the other hand, the Europeans have invested in a machine, which is, roughly speaking, about a third as powerful as the Superconducting Super Collider would have been. This device is called the Large Hadron Collider, which I've mentioned before in these lectures, and the acronym is LHC.

The LHC is in construction right now in the year 2005, with a scheduled end date for construction to be 2007. At some point after that time, the machine will be commissioned, that is, turned on. Typically with such complicated technology, there must be a period of a sort of shakedown, when you must verify that the machine is working as you designed it, so that's one period. Then you have to put detecting devices into the accelerator in order to look for these tiny particles. It is perhaps ironic that to look for the tiniest things, you actually have to spend more and more money, but that's, in fact, the case, where we need more and more advanced technology to look at these smaller and smaller objects.

At this stage, our nation has made an investment. We have spent approximately $500 million to put detecting devices in the Large Hadron Collider. So in this way, we Americans will participate in the search for what lies beyond the Standard Model. Many physicists are convinced that supersymmetry is one of the things that will lie beyond the Standard Model, and that the LHC will be the place where supersymmetry is found, but it's not guaranteed. In the year of 1980, I was visiting the campus of Berkeley in California. There I encountered one of the physicists who introduced the idea of supersymmetry in the Western literature. When supersymmetry began, the Cold War was still underway. Therefore, this topic actually has a bifurcated history, where it was discovered independently by Russian physicists and by Western physicists. The

Western physicists who made this discovery were people by the names of Bruno Zumino and Julius Wess.

In 1980, I was visiting California, and I made a stop at Berkeley. Professor Zumino was a faculty member there. Why did I pick the year 1980? Well, 1980 was the first year that this mathematical idea of supersymmetry was applied to the Standard Model. Prior to that point, theoretical physicists had been studying this mathematics, but they had not applied it to the Standard Model. There were some papers written that year by two independent groups, one of them at Northeastern University. They were led by physicists whose last names are Arnowitt, Chamseddine and Nath. Then there was a separate paper written by a group of physicists in Europe. The second group included the Italian physicist Ferrara, who again was one of the discoverers of the mathematics of supersymmetry. These papers were the first time that people said, "Let's try to embed the Standard Model in a supersymmetric context." These papers were the first ones that suggested the mirror set of particles that we saw on the transparencies.

There was a lot of excitement in the physics community. On encountering Professor Zumino, I said to him, "Gee, isn't this a great development?" The mathematics that a number of us had been involved in for a number of years was suddenly looking as though it was going to become physics. To my very great surprise, this man who was one of the fathers of this idea, looked at me and said, "Maybe." Sometimes ideas in physics turn out to be correct, but not in the way that we think they are. In the Standard Model, we have talked many times about the gluons, the intermediate vector bosons. There's a very special set of equations to describe these things. They're called the Yang-Mills Equations, named after the two physicists who invented these equations in 1954.

However, when they invented these equations, they were actually trying to understand forms of matter that are related to protons and neutrons. It turns out that these equations, applied to protons and neutrons are miserable failures. Yet the mathematics that comes out of this attempt is so potent, and so intrinsically correct, that some 20 years later it was used to describe the weak intermediate vector bosons in a series of papers that physicists such as Glashow, Weinberg and Salam wrote. About ten years after these theoretical papers were written using this mathematics, these objects were found in the laboratory. So even though the initial attempt to apply this mathematics was wrong, the mathematics ultimately came back and described something that we've seen in the world.

It is completely possible that supersymmetry might follow this route. This is why Professor Zumino had some skepticism. Even though many people who have studied supersymmetry find its mathematics exceedingly elegant and exceedingly convincing, until we see it in the laboratory, we will not know if this mathematics is a description of our world. It is possible that this mathematics is being applied in an incorrect way, precisely in the way that the Yang-Mills Equations were applied when they were first invented. If that should be the case, I'm confident, at some point in the future—30 years, 100 years—these equations will come back and be found to describe something else in nature of very great importance.

Lecture Nineteen
Do-See-Do and Swing Your Superpartner—Part II

Scope:

At the end of the last lecture, we saw the possibility that the present understanding of the particle physics population could double. We considered a graphic of a balanced world, in which every matter particle seen in the laboratory has a superpartner yet to be seen. Ideally, seeing these superpartners in the laboratory would be preferable, but it is much more likely that an indirect method will yield the first experimental signatures. This lecture introduces this slightly different way to find the superpartners: through the unification of forces described by the running of the coupling constants.

Outline

I. A familiar statement from high school science states: The formula for the force between two charged objects is the product of the charges divided by the square of the distance.

   A. This equation also comes from Feynman's picture-making method of understanding physics. A picture seen many times in this course shows a photon traveling between two electrons. There is an interesting feature about the mathematics of this picture. Each one of the factors of charge in the standard formula for force law is associated with the points in the diagram at which the electrons and the photon touch. The value for this mathematical feature is called the *coupling constant*.

   B. What would happen in the picture if the charge of the electron were set to 0? The wavy line in the middle of our diagram, representing the photon, no longer touches the v-shaped lines representing the paths of the electrons. If that were the case, there would be no force, and this can be seen in this mathematical formula: $F = \dfrac{\text{charge}^2}{\text{distance}^2}$; if charge is set to 0, then there is no force. A coupling constant controls how strongly the forces repel or attract.

C. In a famous story, Archimedes was sitting in a bathtub one day and had a dazzling thought, causing him to run out of his house and yell, "Eureka!" He had figured out why objects float. His discovery is called *Archimedes's principle*: When an object is placed in water, the amount of force with which the water pushes up is equal to the acceleration of gravity times the density of the water times the volume of water displaced when the object is pushed underwater.

D. A swim in the Great Salt Lake of Utah will show that a person floats at a higher level in that water than would occur in the ocean, because the salt in the lake causes the lake water density to be greater than the seawater density. Archimedes's law says that the buoyant force is proportional to the density. As the density of water increases, the upward force increases. If the density is set to zero, there is no buoyant force. That is like setting the coupling constant to zero in the Feynman diagram.

E. In 1953, two physicists, Ernst Stückelberg and André Petermann, found that the electric coupling constant in the quantum world acts exactly like the density of water in Archimedes's principle. The strength with which two objects in the quantum world repel each other depends on the energy at which the observation is made! The more complicated pictures of the quantum world, when added to the classical picture, change the way in which the photon communicates with a charged particle. With higher energy of the photon or the electrons, the coupling constant changes. Energy is, in the quantum world, as salt is in the water for buoyancy.

F. This is true not just for the electromagnetic force but for all four forces, each of which is characterized by a coupling constant. For the force of gravity, that coupling constant is Newton's constant G; for the weak force, the constant is g_w; for the strong force, it is g_s. Their role is the same as the coupling constant for the electromagnetic force.

II. In the diagram showing the exchange of a message carrier, the electrons may be replaced with quarks. How do quarks interact in this scenario?

A. Quarks interact by exchanging gluons, which keep the quarks permanently bound in the interiors of protons and neutrons. Since the force caused by the gluons has a coupling constant, we can again ask what happens mathematically when the coupling constant is set to zero. As in the case of electromagnetism, the exterior lines in the diagram representing quarks no longer send message carriers back and forth, and without the message carriers, there are no forces.

B. The same is true for gravity. If we replace the electrons in the original Feynman diagram with the Earth and the Moon, the exchange of gravitons is responsible for the attraction between the two. If Newton's constant is set to zero, then the graviton does not carry the message from the Earth to the Moon, and there is no attraction. This is a universal way of understanding how forces arise in the quantum world.

C. The world of Stückelberg and Petermann implies that the coupling constants of nature aren't constant, a surprising result, but a quantum mechanical one—not something that affects us in everyday life. If the charge of electrons in the everyday world were to change arbitrarily, electricity or means of communication would be unpredictable.

III. In 1974, a second chapter was added to this story. In that year, three physicists, Howard Georgi, Helen Quinn, and Steven Weinberg, showed that the rates at which the coupling constants change can be calculated. Their equations describe what is called the *running of the coupling constants*.

A. Plots of running coupling constants can be made on a graph. One axis is the energy axis, labeled E^2, and the other axis measures the strength of the coupling constants: G for gravity, E for the electromagnetic force, g_s for the strong force, and g_w for the weak force. Four points in the graph represent the four forces; gravity is the weakest force, followed by electromagnetism, the weak nuclear force, and the strong nuclear force.

B. The speed to which particles are accelerated determines the energies at which an experiment is performed. Higher speeds imply higher energies. We can illustrate this point by imagining colliding two beams of electrons.

1. The force of repulsion can be measured while the electrons are moving at a relatively small velocity. The same measurement can be made for increased velocities.
2. Remarkably enough, results of such experiments in the quantum world show that the coupling constants do change, as seen in the plot: The strong coupling constant is a bit smaller, the weak coupling constant is about the same, and the electromagnetic and gravitational coupling constants have both increased.

C. What happens at even higher energies? The coupling constants move even closer to each other. Gravity gets even stronger, the electromagnetic interaction stays about the same, the weak interaction gets a little bit stronger, and the strong interaction decreases. At even higher energies, the electromagnetic and weak coupling constants merge. This energy is then called the *electro-weak unification energy*. This change in the coupling constants is actually something that can be seen in a laboratory experiment.

D. At even higher energies, the strong nuclear force, weak nuclear force, and electromagnetic forces all join and have the same strength. The only outlier is the force of gravity. Of course, at an even higher energy, gravity will join the other three forces. This energy is very close to what is called the *Planck energy*, and at this level, gravity and the other forces are indistinguishable.

IV. At this point in the presentation, the discussion has shown how coupling constants *run* in the microcosm. This has interesting implications for the standard model.

A. In a second animated plot of the forces, we see the paths that the coupling constants for each force follow as they become unified. The paths shown can be calculated using the method of Georgi, Quinn, and Weinberg.

B. Notice that the electro-weak unification occurred before the strong force joined. The coupling constant for gravity unifies with the others even later. It is interesting to reconsider this pattern of unification but in the context of a supersymmetric version of the standard model.

C. If the particles in the standard model have superpartners, that has an impact on the running of the coupling constants. Considering a Feynman diagram of the electromagnetic force, the superpartners also could play a role, via loops, of modifying the effects of the standard force carriers. Since superpartners have slightly different properties from ordinary matter, they will change the way that the coupling constant runs in a slightly different way. This makes it possible to ask: If our world is supersymmetric, how do the coupling constants run?

D. The physicist who first performed these calculations, in 1984, was a Russian named Dmitri Kazakov. His work showed that, in the supersymmetrical model, the strong, weak, and electromagnetic forces all unify at the same energy. That's a different behavior than seen in the standard model without superpartners.

E. Physics is often driven by a sense of elegance. In this case, many physicists have argued that the supersymmetrical picture of the unification of forces is a more elegant result than the standard model. Unfortunately, our sense of beauty does not determine physics. Physics is determined by laboratory experiments, but this understanding of the unification of forces offers another window through which to determine whether our world is supersymmetric.

F. In the last lecture, we saw that, depending on the masses of the superpartners, we may not be able to see them for hundreds of years. The coupling constants offer a different path to discovery. The ability to measure the coupling constants does not depend on first seeing the superpartners. Instead, it relies on careful measurement of ordinary forces between ordinary matter. Another way to detect the presence of superpartners, then, is to measure the paths of the coupling constants.

G. If experiments at the LHC indicate that the paths of the coupling constants are the ones we associate with the standard model, this will be damning evidence against the existence of supersymmetry. If the paths of the coupling constants point to a single unification of the strong, weak, and electromagnetic forces, that will be pretty strong evidence that ours is a supersymmetrical universe, even though superpartners are not yet seen.

V. Why should the existence of superpartners matter?

A. The concept of an electron began as an idea. Of course, now, a great deal of present-day technology is based on this idea. This provides a good example of what can happen when humanity discovers new forms of matter and energy. There is a long-term potential for new and unimagined forms of technology.

B. What might these technologies be? Science fiction permits speculation. Perhaps it is possible to transmit ordinary electrons into selectrons. We have seen that, if it were possible to turn off the charges for ordinary electrons, they would still repel each other. But selectrons are bosons; if their charges are turned off, they can pass right through each other.

1. In present-day experiments, a process called *quantum teleportation* is being studied in laboratories. This process is possible because photons are bosons, so many can be superimposed at the same place at the same time.

2. In turn, this leads to *entanglement*, which is also being explored for the possibility to create quantum computers. Since selectrons are bosons, it should be possible to entangle them. To turn off their electric charges would mean something akin to their further transmutation into electron sneutrinos. If these were to accelerate to near the speed of light, their lifetimes would increase while they would retain the attributes of quantum entanglement.

3. Further, if all this were to be a reversible process, the end result would be a matter transportation process, right out of *Star Trek*! While this is all simple and wild speculation, it is interesting to imagine how the discovery of supersymmetry in our universe could be the key to creating transporters.

Readings:

Gell-Mann, *The Quark and the Jaguar: Adventures in the Simple and the Complex.*

Lederman, *The God Particle*.

Lederman and Hill, *Symmetry and the Beautiful Universe*.

Kane, *Supersymmetry: Unraveling the Ultimate Laws of Nature*.

Oerter, *The Theory of Almost Everything: The Standard Model, the Unsung Triumph of Modern Physics*.

Zee, *Fearful Symmetry: The Search for Beauty in Modern Physics*.

Questions to Consider:

1. Describe the running of the coupling constant in the standard model.
2. Describe the running of the coupling constant in the quantum world.

Lecture Nineteen—Transcript
Do-See-Do and Swing Your Superpartner—Part II

At the end of the last lecture, I talked about the possibility that we've seen about half of our world in the course of studying particle physics. We put a transparency up of a very balanced world, where for every matter particle that we've seen in the laboratory, there is a superpartner that we have yet to see. I should be a little bit more specific, however, when it comes to the Higgs Particle. Because if supersymmetry is realized in our world, in fact, there are two Higgs Particles, and that's kind of an unusual feature. When the idea of supersymmetry was first presented, it was a wild speculation. There were very few physicists who took the idea seriously. I was a graduate student when this happened, so the middle 1970s and I was working toward a doctoral degree. In my third year, I came across a research paper by Wess and Zumino, where I first saw the mathematics of supersymmetry.

I understood pretty quickly that this mathematics described the possibilities of new forms of matter and energy that no one had ever thought about. I was very excited. I ran around the Center for Theoretical Physics at MIT, where I was a student, trying to find out if any of my faculty could help me learn this new subject. None of them could. In fact, I was the only person at MIT in those days who had any interest in the idea of supersymmetry, so I persevered. I had an advisor who helped me learn the material. After completing some work where I tried to combine the idea of supersymmetry with the equations of Maxwell, I set my sights on trying to next study supersymmetry combined with the ideas of Einstein. In 1977, I wrote the first Ph.D. thesis at MIT on the subject of supersymmetry. Fortunately, it was not the last.

The other thing that we want to talk about today is a slightly different way to find the superpartners. We would like to, of course, see them in the laboratory. That's the most direct way that we know that this mathematics is an accurate description of our world. There is, however, an indirect way that we can know this, and that's what I want to talk about today, because it is much more likely that it will be this indirect method that provides the first evidence in an experiment that we live in a supersymmetrical world. Again, let me remind you of something you know. Let's go back to our high school science class. If we have two charged objects, and we want to know the force between the two, then we learn in high school science that the force goes like the product of the charges divided by the square of the distance. If the charges are electrons, then we have to use this incredibly small unit, the electron charge.

This equation we have learned also arises by using Feynman's picture-making method to understand physics. Let's think about these pictures. We've seen it many times, where we have our particles being exchanged, with the photon traveling between the two and the electrons in the exterior of the diagram. Here's our familiar picture. Now there's something rather interesting about the mathematics of this picture. Remember, in the high school science class, we were told the force goes like the product of the charge divided by the square of the distance. Each one of the factors of the charge is associated where the electron lines and the photon lines are associated, where they touch each other. There are two places in this diagram where it occurs. On the left-hand side, as the electron first makes its V in its path, we should think about well, gee, the electron charge is at this point. On the right-hand side of the diagram, the other electron makes its turn, and so that's also associated with the factor of the electron charge.

Now let's do something interesting. Suppose we set the electron charge to zero. Remember, it's a mathematical expression. We can do whatever we want to with the mathematics. Of course, we can't do that in the real world. But let's, in the mathematics, set these coupling constants, as they are called, to zero. What would happen to our picture? Something very interesting would happen. We would find that the wiggly line in the middle no longer touches the V-shaped lines on the exterior of the diagram. That would mean, of course, that there would be no force, but that's exactly right. If the force law goes like F is equal to the charge squared over the distance squared, and you set the charge to zero, then there's no force. In the quantum world, it's the message carrier that's responsible for the force. Therefore, when you set the charge to zero, you stop the message carrier from carrying its message. That's a consistent view.

The coupling constant controls how strongly the force is repelled or attracted. If we have like charges, they repel; unlike charges attract. There's a story, of course, that's rather famous about Archimedes. The story says that he was sitting in a bathtub one day and jumped out of the tub because of a certain thought and ran out of his house yelling "Eureka!" Now I don't know if this is a true story, but it's one that one often hears. One of the interesting things about Archimedes is recently we have found evidence that he, not Newton and Leibniz, may well have been the first inventor of calculus. It's a very interesting story, but I'm sure that's for another course.

Why did Archimedes jump up? Well, the idea was that he had figured out why objects float. That's something we now call the Archimedes Principle. What it basically says is that when you put an object into water, the amount of force with which the water presses up is equal to something we call the acceleration of gravity—the difference in the density of the object that you put in the water and the water itself and the volume that is displaced when you push the object under water. That's the Archimedes Principle. Now we're going to use the Archimedes Principle to think about something slightly different. If you were to go and float in the ocean, you might make a measurement about how far you sink into the water. On the other hand, if you were to go to the Great Salt Lake and simply lie out and float on that body of water, you would find that you float at a higher level. The reason for this, as many people know, is that the Great Salt Lake has lots and lots of salt dissolved in the water. What that causes is for the density of the liquid there to be greater than the density of the liquid in seawater.

Remember, the Archimedes Principle says that the buoyant force is proportional to the density, so if the density of the liquid increases, it means that the force upward increases, and therefore you sink less. That's why you float at a higher level. In this example that we're talking about, we can actually see this behavior of what we call the coupling constant in our previous talk about electrons. Let's imagine we could set the density difference to zero. If that's the case, then there is no buoyant force. That's just like setting the coupling constants to zero in our Feynman diagram. In this physical situation, we can see that the amount of force that's applied by the liquid depends on the density, and there are things that you can do to change the density of the water.

That's actually an interesting observation because, in 1953, two physicists, Stückelberg and Petermann, found that the electric coupling constant in the quantum world acts exactly like the density of water in The Archimedes Principle. In other words, the strength with which two objects repel each other actually depends on the energy at which you make the observation. That's like adding salt to the water, a very surprising result. The quantum denizens—because when you actually try to figure out what's going on in terms of pictures, what you essentially find is that the more complicated pictures of the quantum world—when added to the classical picture, change the way with which the electron communicates with the charged particle. The higher the energy of either the photon or the electrons to which the photons are communicating, then you find out that the coupling constant effectively changes. In other words, energy is in the quantum world as salt is in the water for the Archimedes Principle.

Now this turns out to be true not just for the electromagnetic force, because we have learned in our discussion that there are four fundamental forces in the world. There is the force of gravity. There is the weak nuclear force. There is the force of electromagnetism, and there's the strong nuclear force. Each one of these forces is characterized by a coupling constant. For the force of gravity, that coupling constant is Newton's constant, G. For the other two forces, we introduce different constants that are sometimes called g_w, which means the coupling constant for the weak force, or g_s for the coupling constant for the strong force. But their role is to play exactly the same role that we saw for the electromagnetic force. By that, I mean the following. In our famous V diagram with the exchange of a message carrier in the middle, instead of saying that the outside particles are electrons, we can replace them by, for example, quarks.

How do quarks interact? Quarks interact by exchanging gluons. That's what keeps them permanently bound in the interior of objects like protons and neutrons. So if they have a coupling constant, once again you can imagine mathematically what happens when I set the coupling constant to zero. Then what you find is, just as in the case of electromagnetism, the exterior lines, which represent the quarks, no longer send message carriers back and forth. Without the message carriers, there are no forces. This is true for gravity. Again, we're taking our same V-shaped diagram, which is called a Feynman diagram, which I should probably technically say, if we take the Feynman diagram and imagine that one leg is the Earth, the other leg is the Moon, and we're trying to understand the attraction of the Earth to the Moon, then it's the exchange of gravitons that are responsible for that.

If we set Newton's constant to zero, then the graviton does not carry the message from the Earth to the Moon, and therefore there's no longer any attraction. So this is a universal way of understanding how forces arise in the quantum world. Of course, this also applies if we think about the weak nuclear force. We learn that the coupling constants of nature aren't constant, a very surprising result. But it's a quantum mechanical result, not something that affects us in everyday life. Thank goodness it doesn't because, you see, if it were true that the charge on the electron in our everyday experience changed, then, for example, we would have no guarantees about having reliable electricity. We would have no guarantees in having reliable electronic communication techniques, no reliable telephones, no reliable televisions, no reliable radios. So at our level, it's a good thing that the constant doesn't vary.

In 1974, a second part of this story appeared. That year, three physicists, Georgi, Quinn and Weinberg, showed that the rates at which the coupling constants change can be calculated. Now that's a very interesting result because that is—if we talk about our analogy with the buoyant force, you could ask the question—if I add a little bit of salt, by how much bigger does the density change or how much smaller? If I add salt at a faster rate, how does the rate of the density change? There's a correlation between how you change the density, how much salt is added, and how the density is changed. The 1974 result of these three physicists showed that the rate of change of the coupling constants in the microcosmic world are calculable, and they found the equations which describe this process.

Nowadays, what we say is that the equations describe the running of the coupling constant. Let's see if we can illustrate this with a diagram. First of all, we would just want to plot out our coupling constants. We're going to take a plane. One of the axes in the plane, we will label as the energy axis, so we'll use the letter E^2 because we want to measure square energy. The other axis with which we label the plane we'll call the coupling constant axis. That means it's G if it's gravity, it's E if it's the electron, it's g_s if it's the strong force, or it's g_w if it is the weak force. In our world, gravity is the weakest force; this is then followed by the electromagnetic force. Then, a little bit stronger, we have the weak nuclear force—and, finally, the strongest force is the so-called strong nuclear force. Here we see the coupling constant for the strong nuclear force. Here we see the coupling constant for the weak force, here we see the coupling constant for electromagnetism, and here we see the coupling constant for gravity. That's the order in which they occur in our world. The strong force really is strong.

Now let's imagine performing an experiment where we perform the experiment at higher energies. What does that actually mean higher energies? An experiment's energy is defined by how fast we accelerate particles. Remember, energy exists also due to motion; it's called kinetic energy. So the faster an object moves, the more energy it has. You could, for example, imagine taking two electrons and measuring the force of repulsion between them when they're moving fairly slow. You'll have an answer. Now take the same electrons and shoot them at each other with a greater speed. Since they have a greater speed, they have a greater energy. Once again, try to measure the force between the two. Remarkably enough, the results that we have described from the quantum world tell us that the coupling constants change. Here we see them in our image.

If you look to see the coupling constant, you'll notice that the strong coupling constant is actually a little bit smaller than it was when we previously looked at it. The electromagnetic coupling constant is about the same. The weak coupling constant has actually become stronger and, interestingly enough, the gravitational coupling constant has become strong. What happens at even higher energies? In fact, the coupling constants become even closer together. Gravity becomes a little bit stronger. The weak interaction becomes a little bit stronger. The electromagnetic interaction stays about the same, and the strong interaction decreases. In fact, if you look very carefully, you'll see that the electromagnetic and the weak coupling constants look very close together. So what happens at higher energy? They merge.

If you perform an experiment at high enough energy, it is impossible to distinguish between the electromagnetic force and the weak nuclear force. We call this energy the electroweak unification energy. It is mathematically implied precisely by the equations of Yang and Mills. You'll remember I discussed in a previous lecture that there were equations that started out being incorrectly used, but some 20 to 25 years later, we found they actually described the force carriers. Well, in order to produce mass for these things, which is a process which deserves a lecture of its own; it turns out that you must, at some level, regard the forces as having the same strength. This measure of the coupling constants changing is something we've actually seen in the laboratory. Therefore, we know that at higher energies, the electromagnetic force does stay about the same, but the weak nuclear force actually grows.

Suppose we work at even higher energies. Then if you look at our diagram, you'll notice that there's another unification. At this stage, the strong nuclear force, the electromagnetic force and the weak nuclear force have all joined and have the same strength. The only outlier is the force of gravity, which is weaker than the other three. Finally, if we work at an even higher energy, gravity joins the crowd. There's an energy, which is someplace very close to what we call the Planck Energy. The Planck Energy is about the highest energy that we can measure, constructed from fundamental constants. Very close to that level, gravity and the other forces are indistinguishable. If you look at the top of my transparency, you'll notice that I've, in fact, been adding the forces all together. On the far left, we had electromagnetism. Then we had the weak force. Electromagnetism and the weak force join. At a high energy, electromagnetism, the weak force and the strong force. Then finally, as we approach the Planck Energy, all four forces are indistinguishable. That's how coupling constants run in the microcosmic world.

This has very interesting implications for the Standard Model. In fact, let me run this process once more, but this time using some tracks. There we see our coupling constants before, but I have their after images now painted in. You can see that first we achieved electro-weak unification there. Then we achieved electro-weak and strong unification. Then finally, only as we approach the Planck Energy, are all four forces joined together. This set of paths that you see is precisely what was calculated by Weinberg and Quinn and their collaborator, which I mentioned earlier, that the path of the coupling constants can be joined. There's something very interesting about these paths. You will notice that the electro-weak unification occurred before the strong force joined in. Then after the strong force was joined to the other two, gravity joins the crowd.

That provides and interesting mathematical laboratory to think about the supersymmetric version of the Standard Model. Remember, we talked about how the Standard Model has all the particles that we've seen in the laboratory, but they're somehow unbalanced. The world would be a more balanced place if they have superpartners. If there are superpartners, that has an impact on the running of the coupling constants, and it's pretty easy to figure out why this is true. In the pictures that I have shown you, this picture making way of understanding the forces, in the middle parts of the picture, if there are superpartners, they can play that role. Not the standard force carriers, but the superpartners. Since the superpartners have slightly different properties, it means that they will change the way that the coupling constant runs in a slightly different way.

In particular, we can rerun the last image, asking if our world is supersymmetric, how the coupling constants come together. How do they run, in the expression of theoretical physicists? Let's run our movie and answer this question. Let me, before I do that, mention the physicist who actually first did this. It was a Russian physicist by the name of Kazakov, and this calculation was actually done in 1984. Now let's see what the result of the calculation reveal to us. There are our coupling constants that are about the energies that we think of as ordinary energies where we can perform experiments today. Now we're going to imagine doing the experiments at higher and higher energies. Once again, we see that there's a convergence. The strong force becomes weaker, the weak force becomes strong, electromagnetism dies a little bit, and gravity grows a bit stronger.

But the thing that is interesting about the supersymmetrical model is that you will notice when the unification occurs, the strong, weak and electromagnetic forces all unify at the same energy. That's a very different behavior than we saw in the Standard Model without superpartners. In fact, we can go back and make a comparison. Here's the unification in the supersymmetric version. Here is the unification in the Standard Model, which we experimentally have support for. As you can see in the Standard Model version, there is unification first of electromagnetism and weak, and then only later does the strong force join in. Whereas in the supersymmetrical model, all three forces outside of gravity unify at the same place.

Physics is often driven by a sense of elegance, although it is often perhaps very hard to define what we mean by elegance. In this case, many physicists have argued that the picture that is provided to us by the supersymmetrical standard model, where three of the forces join at exactly the same place, is a more elegant result than in the Standard Model, where first you have a joining of two forces, and then the third non-gravitational force joins the crowd. This remarkable result about what the world of supersymmetry is, in some sense, looks pretty, but again, remember physics ultimately is not about our sense of beauty. Our sense of beauty does not determine what we call physics. Physics is determined by a laboratory experiment. Now we have another window with which to find out if our world is supersymmetric.

In my last lecture, I described one way we could discover supersymmetry, which is by actually seeing the superpartners. We've never seen them in the laboratory. Depending on their masses, this may, if we're lucky, occur before 2020. If we're unlucky, we may have to wait hundreds of years in order to actually see a superpartner. In the latter case, of course, the entire time of the debate would be whether this mathematics is physics, or is it philosophy? The coupling constants offer a different path. You see, the ability to measure coupling constants does not depend on us first seeing the superpartners. Instead, what it relies on is a very careful measure of the ordinary forces between ordinary matters as we already understand them. If that's the case, then another way to detect the presence of superpartners—or perhaps something else—is actually to measure the paths that the coupling constants take.

If we perform experiments at the Large Hadron Collider, and they indicate that the paths of the coupling constant are the ones that we associated with the Standard Model, it would be pretty damning evidence against the existence of supersymmetry. On the other hand, if we perform experiments at the Large Hadron Collider, and we find that the paths of the trajectories for the coupling constants point to a single point of unification between the strong, the electromagnetic and the weak coupling constants, that would be fairly strong evidence that we live in a supersymmetrical universe, even though we will not have seen the superpartners. So they have an indirect effect on how the structure of our universe, with ordinary forces and ordinary matter, behaves.

This is a second window, which will be one that will be looked to be exploited at the Large Hadron Collider. As you can see in these plots, you have to be able to work at higher and higher energies to watch the running of the coupling constants, but with that ability, you can plot the paths. You can then make predictions about where they'll meet, and if they all meet in the same place, we can be fairly certain that our universe is a supersymmetrical universe where, even though we have not seen the superpartners, this balance picture that we've constructed mathematically will actually turn out to be a picture of our universe.

Why is this important? The simplest thing I can tell you about why you should care if superpartners exist is to repeat a story of something that happened in the past. I mentioned, perhaps, that the electron was first thought of as an idea. We now, of course, use this idea to base a large part of our technology and our quality of life upon. Whenever we discover new forms of matter and energy, it means potentially there are new forms of technology out there that are waiting for us to exploit them. What might these technologies be? Again, this is in the realm of speculation. I can't tell you exactly, but I can imagine. Suppose it's possible to transform ordinary electrons into selectrons. We know, for example, that if you turn off the charges of electrons, they'll still repel each other. But if you turn off all the charges on selectrons, since they're bosons, they can pass right through each other up to quantum effects.

Imagine a device that changes electrons into selectrons, everything. Then you might, perhaps, have a new way to transport the electrons, that is, change them to selectrons, transport them and change them back. Does that mean perhaps that you might be able to transport them even farther or faster, or perhaps maybe through ordinary matter? We don't know. But some prospect that perhaps looks to us very much like science fiction may well be out there if these new forms of matter and energy exist. That's the importance of new forms of matter and energy.

Lecture Twenty
A Superpartner for Dr. Einstein's Graviton

Scope:

In this lecture, we discover that most of the superparticles will be enormously massive, but one of them, the *lightest supersymmetrical particle* (LSP), will be the least massive. Because this particle is supersymmetrical, it may not function according to the same force laws that act on ordinary matter. In that case, it will be dark and may offer an explanation for the dark matter problem in the structure of galaxy formation.

Outline

I. Research conducted by your professor posited an alternative model for superpartners, but the focus here will be on the version of supersymmetry that has the widest support today.

II. Can we find a mathematically consistent way to introduce mass to the superpartners so they become very heavy while ordinary matter remains very light?

   A. Ordinary matter is paired with supermatter in supersymmetry. If this theory is correct, then the graviton must have a superpartner.

   1. Light can be left-circularly polarized or right-circularly polarized. Physicists refer to these as the degrees of freedom of the photon. This statement is also true for gravitons. The number of degrees of freedom for gravitons, like all force carriers, depends on the number of dimensions being considered.

   2. The formula to find the number of degrees of freedom of a graviton is: $\frac{1}{2} D (D - 3)$, in which D stands for the number of dimensions. For our world, this number is 4. If this is substituted in the formula, the result is 2.

   3. If $D = 3$ in this formula, there are no degrees of freedom for gravity. Thus, if the world were three dimensional (2 spatial and 1 temporal dimension), there would be no gravity, and if there were no gravity, no galaxies, no planets, no stars, no life. The smallest conceivable mathematical construct that results in a force of gravity with degrees of freedom is 4…looks remarkably like our world.

B. Einstein gave us a picture of gravity associated with curvature or acceleration. If a person were placed in an elevator, far from any planet, accelerating upward, all objects, such as a ball, would appear to fall. Even a beam of light would appear to fall because gravity truly does curve light. The message carrier responsible for this is the graviton.

C. For supersymmetry to exist, exchanges of ordinary quanta of matter and energy must be made with superpartners. The object exchanged with the graviton must differ from the graviton by 1/2 unit of spin; it is called the *gravitino*.

D. In the mid-1970s, the equations describing this object interacting consistently with gravity were first written by two groups, Ferrara, Freedman, and van Nieuwenhuizen, and Deser and Zumino. This theory of gravity with a superpartner is called *supergravity*.

E. Supergravity has the same kind of angles associated with it as the photons. Earlier, we said that if the change in two different photons is the same, the result for each is the same electromagnetic field. This statement is also true of gravitons. If the change in two different gravitons is the same, the same gravitational effect is produced.

III. It is thus possible to have a general picture of supergravity, but our world is not supersymmetrical. No doubling of the particles has yet been observed.

A. If superpartners are very massive, they have yet to be seen in experiments. Still, is it possible to find mathematical equations which allow superpartners to be much more massive than the quanta of ordinary matter and energy?

B. Of the known force carriers, only those associated with the weak nuclear force have mass. The other force carriers are massless. This problem was not completely worked out until particle physicists borrowed an idea from physicists who studied magnets.

   1. Mathematical equations can be written to describe a phenomenon called *spontaneous magnetization*, where materials that are not magnets at normal temperatures

become magnets if cooled sufficiently. These same equations were borrowed by particle physicists to study how to include mass for the force carriers.
2. For years, it was known that the most obvious way to include mass for force carriers results in time-like particles. Mathematics used to describe magnets came to the rescue. Brout, Englert, Guralnik, Hagen, Higgs, and Kibble applied equations borrowed from the description of spontaneous magnetization to particle physics, the result being the *Higgs mechanism*, which provides a way to give mass to ordinary force carriers without introducing time-like particles.

IV. A look at how the Higgs mechanism works is in order.
   A. It is believed that a spin-0 particle exists in nature called the *Goldstone particle*, whose energy takes the form of a Mexican hat. Our animation shows a ball, representing the Goldstone boson or bosons, on top of a hill. Below the Mexican hat, the animation has three dots, representing ordinary force carriers, such as the intermediate vector bosons.
   1. When the Higgs boson and the Goldstone fields are at the top of this energy potential, the ordinary force carriers are massless. As the Goldstone and Higgs bosons together roll down this potential, the force carriers grow massive.
   2. In the picture, one of the force carriers, the photon, did not grow mass.
   3. In a different animation, we see that when the Goldstone boson reaches the bottom of the energy potential, it leaves another energy field related to the Higgs boson throughout space and time.
   4. The disks in the second animation represent the remaining energy of the Higgs boson pervading space. As an ordinary particle that would be massless moves through this field, it grows mass.
   5. This seemingly inelegant method is the only known way to provide mass to all matter and force carriers that occur in our universe without introducing other problems like time-like particles.

©2006 The Teaching Company

B. This is how physicists believe mass is given to the ordinary force carriers as well as matter fermions, but remember, there is the other problem of giving mass to the superpartners.
   1. If supersymmetry is a feature of this universe, then the force carrier for gravity must have a superpartner. This superpartner has been identified as the gravitino, which spins at a rate of 3/2. Initial attempts to write equations for this object also indicate an expectation for it to be massless.
   2. Gravity couples to everything, both ordinary matter and supermatter. This means that the superpartner of gravity must also couple to ordinary matter and supermatter. This cannot be all there is to this story. If it were, long ago it should have been possible to detect the effects of a massless gravitino.
C. When bosons acquire mass through the Higgs mechanism, the kinds of angles associated with change in a force carrier more or less disappear, and when these angles disappear, some symmetries are apparently lost. This process is called *spontaneous symmetry breaking*.
D. In the early 1980s, physicists wrote a model in which, initially, everything is massless. To include the effects of gravity and its superpartner, the model repeated for supersymmetry exactly the process that was carried out to give mass to the ordinary force carriers. In other words, the goal was to find a supersymmetrical version of the spontaneous symmetry-breaking mechanism.
E. What are the implications of this model?
   1. In the ordinary Higgs mechanism, as the Goldstone bosons roll down the energy hill, other particles gain mass. It is necessary to repeat that process in the context of supersymmetry.
   2. In supersymmetry, there should be an expectation to find a super-partner of the Goldstone particle sitting at the top of an energy hill called a superpotential. If the ordinary particle that sits at the top of a potential is spin 0, its superpartner would have spin 1/2. This supersymmetrical partner has been named the *Goldstino*.

3. In the supersymmetrical picture of the Higgs mechanism, there is both ordinary matter and supermatter. Also at the top of the hill picture is the Goldstino. As the Goldstino rolls down the Mexican hat, the ordinary particles remain massless (their masses are induced by the ordinary Higgs boson), but their superpartners all become much more massive, including the gravitino.
4. Thus, by simply writing the equations in which gravity has a superpartner, then repeating exactly the same process that gives mass to the force carriers and matter of our world, all superpartners can be shown to grow incredibly larger masses.
5. This mechanism is now called the *spontaneous breaking of supersymmetry*, and the implication of it is that supermatter is always much, much heavier than ordinary matter.

F. This model solves another problem. As supermatter gets heavier, the superpartner to the graviton grows an incredible mass. This needs to be compatible with observation of our universe as all present-day experiments and observations point to one massless particle that is associated with communicating gravitational forces. If the supersymmetric Higgs mechanism had not worked, then in addition to the graviton, our universe would require a second object, the *gravitino*, responsible for transmitting gravitational forces over long ranges.

V. With this picture, many of the previously discussed problems of supersymmetry disappear. We have never seen superpartners because they are so massive. The Goldstino, which is the superpartner to the graviton, is naturally a very massive object, and when it disappears from the theory, it leaves behind an energy field that gives mass to everything not yet seen with present-day instruments.

Readings:

Greene, *The Elegant Universe* and *The Fabric of the Cosmos: Space, Time and the Texture of Reality*.

Kaku, *Hyperspace: A Scientific Odyssey through Parallel Universes, Time Warps, and the 10th Dimension*.

Kane, *Supersymmetry: Unraveling the Ultimate Laws of Nature.*

Questions to Consider:
1. What aspects of our world might supersymmetry help explain?
2. How does the Higgs mechanism work in our universe and how can it be applied to a supersymmetrical universe?

Lecture Twenty—Transcript
A Superpartner for Dr. Einstein's Graviton

In the last lecture, we were looking, and with some specificity, about supersymmetry. Remember this is, of course, really a course on string theory. So we have to think for a moment, why were we doing that? The answer is that supersymmetry is one of the signatures that string theory almost predicts about our world. If supersymmetry is discovered in the laboratory, it will be an enormous boost for the idea that string theory is an accurate description of nature. Supersymmetry, however, would be good for other reasons. Let me admit that supersymmetry could be present without string theory. Remember, I commented that most string theories produce notes in pairs, and that's what we would think of as supersymmetry.

Supersymmetry can fix other mysterious problems in our universe. In one of my very beginning lectures, I talked about something called dark matter. Dark matter explains some things, but it's a puzzle. For example, the formation of galaxies over the time scales that we need in order to describe our universe doesn't work if we just include the matter that is contained in stars that are visible. We have, in order to get the mathematical description to describe our universe, to assume that there's more matter than we can detect with our telescopes, about nine times more. We seem to live in a universe where ordinary matter is only one-tenth of the matter that's present. One of the things that superpartners provide is a candidate for dark matter. In this other mirror part of the universe that's proposed by supersymmetry, there's an object called the LSP. That stands for the lightest supersymmetric particle. It's some combination of the things that we see on the right-hand side of our table. It's not particular clear which one, but one of them will be the LSP.

Why is the LSP so important? Most of the superparticles, as we will discover in this lecture, will be enormously massive. But one of them, if they exist, has to be the lightest one. The lightest one, it turns out, cannot disappear and cannot turn into anything else. It has to be absolutely stable. That means that if we live in a supersymmetric universe, perhaps all of the higher-mass superpartners have gone away. But if they have gone away, they must have transmuted themselves into this lightest supersymmetrical particle, which means that it would still be here. Since it's supersymmetrical, it might not have the same sorts of force laws that ordinary matter has. In that case, it would effectively be dark. So the LSP is an almost ideal candidate to explain the dark matter problem in the structure of galaxy formation. It would fix a problem if our world were supersymmetric.

The description of superpartners I've given you so far is the predominant one, but it's not the only possibility. A few years ago, working with a colleague by the name of Oleg Lebedev, I studied the mathematical models where the LSP, in fact, was one of the squarks. Now why is that important? Squarks are, of course, related to quarks, but all quarks are bound in the interior of nuclear matter. So if the LSP was actually a squark, it means that you would find the lightest supersymmetrical particle bound inside of a hadron. If that's the case, finding supersymmetry turns out to be much more difficult than the view that I've described previously, where the LSP is not a squark. Now this is very unlikely, but mathematically it's good for us to know of the mischief that nature can make should she choose.

The bulk of this lecture is going to be talking about this mass issue. Remember, I kept saying the superpartners have to be massive. So the question becomes can we find a consistent way in which to introduce mass for the superpartners so that they become very heavy, whereas ordinary matter stays light? It turns out that the answer to this question requires us to do something that we have ignored. We've talked about the fact that supersymmetry says that things come in pairs, that ordinary matter is paired with supermatter. What about gravity? If the principle of supersymmetry is correct, then gravity, the graviton, has to have a superpartner. So let's go back and review some things on gravity. Again, in a previous lecture, we've learned to think about gravity as somehow associated with curvature. Now there's an interesting thing about gravity and the force carrier for gravity.

I described the fact that light has two polarizations. It can be left circularly polarized or right circularly polarized. This statement is approximately true for gravity also, so there are actually two degrees of freedom for gravity. But one of the most interesting things about this observation is that if you're a mathematician and you know how to describe gravity, then you can describe gravity in all dimensions. In particular, one of the peculiar things about gravity is that the number of degrees of freedom that it contains depends on the dimension. In fact, if you look at this expression I've written, I say that the graviton in D dimensions has the number of degrees of freedom that are equal to one-half D times B-3. For us, the number D is four. So you plug that into this formula, and it becomes one-half of four. Then four minus three is one. One half of four is two. Yes, that's exactly the two polarization states of the graviton.

A curious thing about this formula, it says that if D is equal to three, there are no degrees of freedom of gravity. That's actually very interesting because if the world were three-dimensional, there would be no gravity. If there's no gravity, there are no galaxies. If there are no galaxies, there are no planets, no stars and no life. So there's actually a mathematical reason why the smallest conceivable mathematical construct in which you get a force of gravity turns to look remarkably like our world. It happens at D equals four. Now you can have gravity for higher dimensions, but it turns out that four is the lowest dimension in which you could have gravity.

Now, supersymmetry—supersymmetry, I remind you, means that we're allowed to exchange. Going back and using our ladder analogy just one more time, we can think of the blue square as being the graviton. That's the one that appears below in our diagram. But if supersymmetry is correct, then we are allowed to change things and look for something that has similar properties, but here corresponds to the brown square. We must find something like this object for the force of gravity. Einstein gave us a picture of gravity associated with curvature or acceleration, which is a neater way to think about it. How does that actually work? Well, Einstein essentially says that if you were in an elevator that was accelerating upward, then all objects would appear to fall. So a ball would fall, but interestingly enough, this same set of equations said that a light beam would fall. Gravity curves light. That's something we know to be true.

There is some kind of message carrier responsible for all of these actions. It's called the graviton. It's the waves that we saw coming off of our neutron star system many, many lectures ago. By this exchange rule, the graviton, the message for the gravitational force, must be exchanged with another object. This other object has to differ by one-half unit of spin. The graviton spins at a rate of four times the rate of the electron, or spin two. This other object spins at a rate of three-halves. Now it turns out that mathematically, physicists have been wondering about these sorts of things for a very long time. In fact, in 1939, one can find the first paper where two physicists by the names of Fierz and Pauli—Pauli once again; he did a lot of good physics—wrote equations which describe objects which are spin three-halves. So the mathematical tools have been around for a long time.

This spin three-half object has a name also; it's called the gravitino. In the late to middle 1970s, the equations which describe this object interacting consistently with gravity were first written down. It was done by two groups of physicists, Ferrara and van Nieuwenhuizen and Freedman; and then separately there was work by Deser and Zumino. This theory of gravity with this superpartner is called supergravity. In fact, your lecturer has spent a large time of his research career studying such theories. These theories also all have the kinds of angles that we associated with the photon. Remember in the lecture about the forces, we had this one demonstration that had two bowls that were different, but the changes—that is, the number of balls coming out—were the same. The comment was that even if we have two different photons, if the change in the photon is the same, then you get the same electromagnetic field.

This statement turns out to be true about the graviton too. If you have two different gravitons, but the change in them is the same, you produce the same gravitational effect in nature. This also applies to the superpartner. So the theory of supergravity was created to respect this property, and its actual creation date was 1975. We have this picture of supergravity where the superpartner to the graviton is around. But our world's not supersymmetric; that is, we don't see a doubling of the particles. We see electrons, but we don't see selectrons. Why not? I have argued previously that if supermatter is very much more massive, you can't see it because we don't have the technology yet. But is it possible to find a mathematical equation that allows supermatter to be very much more massive? In fact, the problem of mass in these theories is much more complicated than you might imagine.

Remember, without gravity, we have the strong, weak and electromagnetic forces. Of the force carriers, only the force carriers for the weak force have mass. The others are all massless. If we consider the force carriers, some have mass, some don't. The ability to give mass to force carriers is a complicated problem. In fact, so complicated that for many years, if you looked at the physics literature, there was no consistent solution to this until particle physicists borrowed an idea from people studying magnets. In the mathematical descriptions of magnets, you can write an equation which mimics something that happens in nature. The something is called spontaneous magnetization. There are materials that, at normal temperatures, are not magnets, but if you cool them sufficiently, they turn into magnets. This is a phenomenon that's been known for a very long time to physicists, and by studying these kinds of materials, sets of equations were developed.

These same equations were borrowed by people studying particle physics who were worried about including mass for the force carriers. In particular, if you want to introduce mass for force carriers, if you try the most obvious thing, you introduce time-like ghosts. My goodness, we've talked about time-like particles before. Often, we also refer to them as time-like ghosts. Even worse sorts of ghosts appear if you just try to give mass to a force carrier of the type similar to the photon or the gluon. So there was a point at which things were stuck. This is where magnets came to the rescue. By taking the equations for magnets that form when you get spontaneous magnetization, a group of physicists deduced that these same equations could introduce mass for the force carriers. It's a fairly large group of physicists who made contributions. These include Brout, Englert, Guralnik, Hagen, Higgs and Kibble, who all worked on this problem in the 1960s.

We now refer to the result of their work as the Higgs Mechanism. The Higgs Mechanism turns out to be a way to give mass to ordinary force carriers. Now let me try to explain the Higgs Mechanism. Again, rather than to bore you with equations, we're going to do this with pictures. The Higgs Mechanism works as follows. It says that in nature you have to have a particle, the so-called Goldstone particle, whose energy takes the form of what we call the Mexican hat. Here we have a picture of that. If you look at the image, you'll notice there's a ball on the top of this hill. This ball represents the Higgs and, more importantly, Goldstone bosons. It's a spin zero particle in nature whose energy has the shape of this object we call the Mexican Hat. Below this object, you will notice that there are three other dots. These other dots represent ordinary force carriers like the intermediate vector bosons.

The Higgs Mechanism says the following—that when the equations describe this situation that we see at the beginning where the Higgs boson and the Goldstone field are at the top of this energy potential, the ordinary force carriers are massless. So the intermediate vector boson has no mass—neither the charged one, the neutral one, nor the negatively charged one. However, as the Higgs and Goldstone bosons together roll down this potential, you'll notice at the bottom of my diagram that the objects are growing more massive. In fact, this is the great discovery that now bears the name of the Glashow-Weinberg-Salam Mechanism. They wrote the equations which prove that this picture that we have applies to the force carriers in our world.

You'll notice in our picture one of our balls didn't grow in mass. The reason is actually quite important. The one that doesn't get mass is the photon. We can write equations where the intermediate vector bosons and the photon initially were all massless. But as the Higgs field rolls down the bottom of its potential, and the Goldstone field in particular rolls down to the bottom of its potential, we get mass. Now you might wonder how this works. Again, I can't show you the mathematics, but I can give you a sense of what goes on. You see, when the Goldstone field goes to the bottom of that potential, it leaves an energy field throughout all space and time, which acts somewhat like syrup. Therefore, we have the following picture. These discs that you see here represent the remaining energy from the Higgs field pervading all the space. Now as an ordinary particle that would be massless moves through these planes, this energy field, the Higgs field, you'll notice it is growing mass.

We think that this is the mechanism by which all mass occurs in our universe. In fact, Nobel laureate Leon Lederman has written a book called *The God Particle*, and it's about the Higgs boson and its role in providing mass in the universe. So that's how we give mass to force carriers. But remember, the problem that we're actually after is not just giving mass to the force carriers. We want to give mass to the superpartners, so we have to worry about how to do that. We have to go back a bit now, because remember I said that if supersymmetry is true, then the force carrier for gravity must have a superpartner. We've actually identified this superpartner as an object that spins at a rate of three-halves, so we know how to write these equations.

When you write such equations, what you find is that this spin three-half object also is a massless object. Since it is a message carrier for gravity, gravity couples to everything. Gravity couples to ordinary matter; gravity must couple to supermatter. That means in particular that the superpartner of gravity must couple to ordinary matter. As well, this superpartner to gravity must couple to supermatter; both of them. That's what supersymmetry means, right? It means balance. They both have to appear in a balanced manner. These equations are pretty simple to write out at some level, and one can do so. Now what we want to do is to use a trick. A moment ago, I described the process by which the weak intermediate boson that's neutral, the charged intermediate boson with a positive charge, and the negatively charged intermediate boson all grow mass through the Higgs Mechanism.

When you grow mass through the Higgs Mechanism, some interesting things go on. One of these things that goes on is that the kinds of angles that are associated with change of a force-carrying particle also more or less disappear. Symmetries disappear. So when these angles go away, it means certain balances are lost, or at least apparently lost. We call this entire process spontaneous symmetry breaking. What we now want to do as a strategy—and this is what was done in these early papers in the 1980s—was to write a model where initially everything is massless, but you want to include the effects of gravity, as well as its superpartner. The next step is you want to repeat for the supersymmetry exactly the process that was carried out to give mass to the ordinary force carriers. In other words, you want to find a supersymmetric version of the spontaneous symmetry breaking mechanism. In the early 1980s, two groups of physicists independently figured out how to write such equations.

So you write these equations, and then you ask a very simple question. What is their implication? It goes as follows. Remember, in the ordinary Higgs Mechanism, there was a particle that we refer to as a Goldstone boson that sits at the top of an energy hill. Then as it goes down the hill, other particles gain mass because it leaves behind a field that pervades all of space. That was those discs we saw. We want to repeat that now, but in the context of supersymmetry. How does that work? Well, supersymmetry always means balance. So if there's a Goldstone particle that sits at the top of an energy potential, by symmetry, you would expect to find a superpartner that sits at the top of a superpotential. If the ordinary particle that sat at the top of the potential was spin zero, its superpartner that would sit at the top of a potential would have spin one-half, because we always differ in superpartners by one-half a unit of spin.

This supersymmetric partner sitting at the top of the potential is named the Goldstino, so it's related to the Goldstone field, which was the boson that introduced mass. Remember we're all talking about mathematics here. So you write the equations that describe the words that I have given, and then you look at the equations and solve for their implications. Let's look at a picture to see what happens. Here's a picture we see. At the bottom of the picture, we have both ordinary matter as well as supermatter. At the top of this picture, we have the Goldstino, not the Goldstone, boson. As the Goldstino rolls down this so-called Mexican hat potential, you will notice that our ordinary particles, which we had illustrated as small balls along the bottom of the diagram, have grown mass. The thing that's really interesting about the pattern of the growth of mass is the large balls that you see in those pictures are all superpartners. They are squarks, they are selectrons, they're Higgsinos, there're photinos, and there are the gluinos. In other words, they are the superpartners to our ordinary matter.

The smaller balls that you see in this diagram, representing the results of an equation—the smaller balls are, in fact, ordinary matter. They're electrons, they're muons, they're tau particles, and they're affiliated neutrinos, and they're the quarks, as well as the ordinary force carriers. It turns out that almost by magic, by simply writing the equations where gravity has a superpartner called the gravitino, and then repeating exactly the process that gives mass to the force carriers of our world by introducing a Goldstino, all of the superpartners in a very natural manner, without sort of excessive manipulation of the equations, can be shown to grow masses. This was the great discovery that excited the particle physics community in the early 1980s—that there was a mechanism that we now call the spontaneous breaking of supersymmetry, which we've seen illustrated in this diagram, whereby the implementation of this mechanism has the implication that supermatter is always much, much heavier than ordinary matter.

This actually solves another problem because, as supermatter gets heavier, the superpartner to the graviton grows mass. Of course, that's exactly what we need in our universe because, you see, in our universe, as far as we can tell from our experiments, there is one massless particle that's associated with communicating gravitational forces. If the supersymmetric Higgs Mechanism had not worked, then in addition to the graviton transmitting gravitational forces back and forth, there would have been a second object. The gravitino would have been responsible also for communicating gravitational forces. Since gravitational forces are long-range, those tend to be forces that one could very easily detect; not by necessarily writing down equations for the quantum world, but instead by looking up at the galaxies, looking at the solar system, looking at the structure of the universe on the large. Because remember, although gravity is the weakest force, it controls the large-scale structure of the universe.

If the superpartner to the graviton had not grown mass, we would have very easily been able to detect that by the structure of the universe itself. We wouldn't need particle accelerators. So we have a coherent picture here, a picture that most of the physics community finds compelling, but it is, of course, still one which needs experimental verification. As you can see, things that look as if they might be problems have almost virtually disappeared from this model. We understand why we've never seen superpartners; because they're so massive. They are so massive because the ghostino, which is the superpartner to the graviton, is a naturally very massive object. When it disappears from the theory, it leaves behind an energy field. It gives mass to everything that we don't see in the laboratory.

How much mass? That's the real question and we don't know the answer. In my previous lecture, I talked about the fact that in order for this balanced supersymmetric view of the universe to be consistent with the experiment, all the superpartners have to be more massive. I even put some limits on how much more massive. We think that they're somewhere between 300 and 1,000 times as heavy as ordinary matter. If they're any heavier than that, other theoretical difficulties appear. So we think that certainly within 300 years or so, assuming continued human progress, we would actually expect to produce a superpartner. Again, in a previous lecture, however, I talked about the fact that our force detection will likely occur by looking at one of the coupling constants.

The thing which is perhaps most surprising about what I've just described, at least by those of us who study the mathematics is, as I said earlier, with almost no manipulation of the equations, we arrive at a picture of the universe that is consistent with supersymmetry. It has the superpartners, but they're so massive that they have avoided our present-day technology. For many physicists, this is extremely satisfying. Is it physics? I don't know. If we detect superpartners, we haven't detected strings, but it's an enormous boost for string theory.

Lecture Twenty-One
Can 4D Forces (without Gravity) Love Strings?

Scope:

String theory did not begin as a project to describe the universe. In the early days of string theory, around 1968, physicists were simply looking for a way to understand nuclear matter. What we are now finding is that string theory can help us describe a physical phenomenon, the running of the coupling constant, that we have not yet been able to calculate mathematically. In this lecture, we will look at this phenomenon, along with other new ideas in string theory, including the *KLT relations*, *anti-de Sitter space*, and *conformal field theory* (AdS-CFT), the *brane world scenario*, and *NSR theory*, that may offer answers to real-world questions about the workings of nuclear matter.

Outline

I. The lecture starts with a representation of a proton, consisting of a bubble and, on the inside, two up quarks and a down quark.

   A. Nuclear matter comes in a second form, in objects called mesons. A representation of a meson has a quark and an anti-quark in its interior.

   B. In 1968, physicists were struggling to determine how nuclear matter was put together. One possible mathematical description was called the *dual resonance model*, presented by Veneziano. The competing model was the *relativistic constituent model*, which ultimately, was judged to be the more accurate model.

   C. The dual resonance model went on to become string theory, which can potentially be used as a way to understand the entire universe.

II. As mentioned, strings come in two varieties, open and closed.

   A. Among the closed strings, there is always a mode, or a manner of vibrating, that has the properties of the particle that carries the force of gravity, the graviton. The open string always has a mode that has the properties of a photon-like force carrier.

B. Strings seem naturally to exist in higher dimensions. When open strings were first studied, physicists thought that they must describe something similar to Maxwell's equations but in 9 spatial dimensions and 1 temporal dimension (for the superstrings) or 25 spatial dimensions and 1 temporal dimension (for the bosonic string).

C. To understand completely the mathematics of such objects, we must be able to count the number of degrees of freedom. The number of degrees of freedom for a 10-dimensional photon is different than that for a 4-dimensional photon; in general, the formula for determining the number of degrees of freedom is $D - 8$ for photon-like objects.

D. An open string carries charges at its ends, while a heterotic string allows charges to be distributed along the length of the string. In 1969, the only method for putting charge on the ends of the string relied on work of Chan and Paton; these charges are called *Chan-Paton factors*.

E. The charges on the ends of the string are not limited to electrical charges. The weak nuclear charge or the strong nuclear charge can also be carried on the ends of the string. However, if the charge is something other than the electrical charge, then the particle that carries the charge must be the appropriate one. For example, if the charge at the ends of the string is color charge, then instead of a photon, the object that carries the charge must have properties like the gluon, which keeps quarks bound to the interior of nuclear matter.

F. Much current research in string theory is devoted to understanding how protons and other nuclear particles work.

III. It is helpful to look at a mystery surrounding these particles.

A. A return to the graph showing the strengths of the four force carriers is a convenient starting point. In looking at this plot in relation to variations with energies, we can see that the coupling constants move in opposite directions. For example, probing an object like the proton at smaller and smaller energies results in the force that couples the quarks together becoming stronger and stronger.

B. In fact, in this version of the graph, the coupling constant for the strong reaction is shooting off into infinity. This phenomenon is required, as presently quarks are thought to be bound to the interior of particles, such as the proton or one of the mesons.

C. Physicists do not know how actually to calculate this behavior of the coupling constant for the strong nuclear force in the described scenario, although there is a good deal of experimental evidence indicating the current understanding is correct. Using experimental probes at higher energies, this coupling constant is found to decrease.

D. The question of why quarks can never escape the interior of hadronic matter reduces to understanding why the coupling constant for the strong force shoots off to infinity at smaller and smaller energies.

IV. In recent years, new ideas have come from string theory to try to understand this strange behavior.

A. Part of this work comes from three physicists working in 1986, Kawai, Lewellen, and Tye, who observed that there must be some very deep, hidden mathematical relationship between a theory of gravity, provided by closed strings, and a theory that includes left-movers and right-movers, which are equivalent to open strings.

B. Earlier animations showed two strings, one only supporting vibrations that moved in a left-hand sense, and one only supporting vibrations moving in a right-hand sense. Each of these has something to do with photons, gluons, or other force carriers in the standard model but not the graviton. This mysterious connection is now called the *KLT relations*.

C. Remember the original problem: Why are quarks the only particles whose force carriers have a coupling constant that runs off into infinity at smaller and smaller energies?
1. Think of a hadron as a bubble with quarks inside, and imagine that you could grab that bubble and start to pull on it.
2. Using Feynman diagrams, the forces between particles are always carried back and forth by message particles, or *gauge particles*. For the strong force, these are the gluons.

3. As the hadron is stretched apart, in our animation, gluons appear, moving back and forth between the quarks.
4. Photons may fly around any path between a positive electrical charge and a negative electrical charge, but gluons cannot. They must remain packed tightly inside the bubble; when gluons are packed together, their energy starts to increase drastically.
5. Einstein taught that energy and mass are the same thing. As the quarks are moved farther apart, the gluons increase their energy to the point at which their energy can be converted into the mass of newly created quarks. In turn, new quarks allow the bubble to split; the result is two particles: one with a quark/anti-quark pair and the other with a three-quark combination, a baryon.
6. No one has ever done a calculation to prove this process works as described above.

D. The KLT relations give us an alternative way to study this process.

V. In 1997, a totally new idea, AdS-CFT, was proposed. AdS stands for *anti-de Sitter space*; CFT stands for *conformal field theory*.

A. In the 1920s, de Sitter was the first person to argue that our universe is one in which the curvature tensor is non-zero and positive. This situation is described as a de Sitter space or dS space. In anti-de Sitter space or AdS space, the curvature is non-zero and negative.

B. In a universe in which the only particles are those like gluons, the only sources of energies are those associated with gluons. However, if the mathematics of a theory that describes only gluons is given, it is found that the physics of the system always looks the same, no matter how deeply we look at it. This is called a *conformal system*. The field in CFT is the field of the gluons, and of course, T stands for theory.

C. AdS-CFT examines the relationship between a system that involves only gluons and relates it to a different theory about gravity with anti-de Sitter geometry, a geometry in which the curvature is negative.

D. This amazing discovery was made by Maldecena, a student of Witten. The discovery marked the beginning of the second string revolution. The first string revolution had occurred after Green and Schwarz discovered the magical number 496 that we talked about earlier.

VI. To explain the shooting up of the coupling constant mathematically, physicists need to exploit new mathematical ideas. Some new mathematics has recently been proposed in the form of objects called *branes*.

A. In an animation showing a small dot in the center of the screen, that dot may be considered to describe a charged particle, such as an electron. If it is an electron, it is free to move about.

B. Instead of calling this electron a dot, let's call it a *0-brane*. If it were a 1-brane, it would look like a line; in fact, it could be one of our strings. We could also construct a 2-brane, which would look like a two-dimensional plane.

C. In our world, the only other such object we could construct would be a 3-brane. In the mathematics of the string, however, nothing stops repeating this process all the way up to 10 as this corresponds to the 10 dimensions of string theory. The collection of all possible branes is called *Dp-branes*.

D. All the branes described must also be able to carry charge. Furthermore, they all must allow for force carriers that can couple to them.

E. Using Dp-branes has allowed us to find mathematical relations that imply a connection between a theory in which there are only gluons and a theory in which there is only gravity, and to derive an equation that connects gravity to a theory that involves only gluons. This discovery also enables physicists to calculate the coupling constant for the strong nuclear force at smaller energies.

VII. Witten has made an even more startling suggestion about how string theory can contribute to our understanding of the strong force.

A. Recall the notion of anti-commuting numbers: The order of multiplication doesn't matter for ordinary numbers, for example, $4\times 3 = 3\times 4$. But in the 1800s, Grassmann studied numbers that had the property in which the order of multiplication does matter.

B. The anti-commuting property leads to a description of string theory called the *Neveu-Schwarz-Ramond (NSR) theory*. This theory includes strands that describe the string, as well as different mathematical strands that obey anti-commuting multiplication rules.

C. For every strand of the string, mirror copies can be introduced that possess the opposite multiplication law. One mirror copy leads to the

N = 1 NSR version of the string; two mirror copies lead to the N = 2 version of the string. From the latter version, mathematical objects emerge that are very similar to spinors, but these are called *twistors*, first proposed by Penrose in the 1960s.

D. The notes of the N = 2 string obey Einstein's hypotenuse only if there are two time dimensions!

VIII. As we can see, a good bit of esoteric mathematics is being marshaled to understand nuclear particles.

A. Although many people assert that string theory cannot be tested, this new horizon of string theory seems to be leading to mathematics that will allow the ability to calculate a physically interesting property, namely, the running of the coupling constant.

B. Even though AdS-CFT has started to yield something that looks like real physics, it's not quite there yet. The only known way to test this idea is to work with something called a *type-IIB superstring*.

C. In the next lecture, we will discuss all five of the superstrings. The IIB string was initially thought to describe only gravity and other denizens (but none like the gluons); with AdS-CFT, there is now evidence that gravity can be connected to the world of gluons. The mathematics of AdS-CFT implies that gluons can have four supersymmetrical partners, that is, four gluinos. This concept is called *N = 4 supersymmetry* and is even harder to accept than the small concept of supersymmetry with only one partner for each particle in our world.

Readings:

Gell-Mann, *The Quark and the Jaguar: Adventures in the Simple and the Complex.*

Lederman, *The God Particle.*

Lederman and Hill, *Symmetry and the Beautiful Universe.*

Oerter, *The Theory of Almost Everything: The Standard Model, the Unsung Triumph of Modern Physics.*

Zee, *Fearful Symmetry: The Search for Beauty in Modern Physics.*

Questions to Consider:

1. Describe the running of the coupling constant as it relates to the strong nuclear force.
2. What are some of the approaches that give physicists alternatives for calculating the coupling constant?

Lecture Twenty-One—Transcript
Can 4D Forces (without Gravity) Love Strings?

String theory did not begin as a project to eat the entire universe. In the early days of string theory, around 1968, the goals were much more modest. Physicists were simply looking for a way to understand nuclear matter. Earlier in our lectures, we visited the quantum world, and in the quantum world, there are many denizens. Among them are the ones that form nuclear matter. Here, for example, we have a representation of a proton. It consists of this bubble; the bubble itself is the actual proton. Then in its interior, the smaller pieces of matter that we call quarks. In this case, it would be a proton. There has to be two up quarks and a down quark. However, the quarks have to be combined in such a way that you take one of each color. Here we've illustrated one red quark, one yellow quark and one blue quark.

Nuclear matter also comes in a second form, obviously they're called *mesons*. This is a representation of a meson. Instead of having three quarks in the interior, you have to have a quark and an anti-quark. So although it's really difficult to tell from this illustration that one of these is an anti-particle, that's what it has to be in order to form a meson. So physicists were struggling with trying to understand how nuclear matter was put together. At the time, one possibility was something called the dual resonance model, presented by Gabriel Veneziano as a possible explanation. The other competing model was called the relativistic constituent model. Ultimately, the relativistic constituent model won out the battle and banished the dual resonance model.

The dual resonance model is actually what went on to become string theory. In using the ideas of string theory, ultimately it was realized that it's much more potent than just trying to understand nuclear matter; that it potentially can be used as a way to understand our universe. Remember when we talk about strings, there are two varieties. There are the kinds that have their ends free, and then there are the kinds that are closed. The closed ones look like little loops. And when we look at the notes or, more technically, the modes of vibrations of these things, we can find one distinguishing feature. Among the closed strings, there's always a mode, which is a manner of vibrating that has the properties from our level of looking like a particle which carries the force of gravity, the graviton. The open string, on the other hand, always has one mode of its way of vibrating that looks similar to a photon.

The thing that's interesting is that we learned many lectures ago that strings seem naturally to want to live in higher dimensions. So when the open strings were first studied, it was thought that these must describe something that's kind of like Maxwell's equation, but in a dimension with nine spatial directions and one temporal direction. If you understand the mathematics of such objects, you can actually count the numbers of degrees of freedom. We've talked about this notion of degrees of freedom before. The number of degrees of freedom for a ten-dimensional photon is different from that of a four-dimensional photon. It's, in fact, in general, the dimension minus eight. String theory fails for hadron physics, but then it comes back and starts to perhaps talk about things that look like electromagnetism.

One of the things about strings, which we've talked about, is that if you have an open string, you can put charges at the end of the string. You have to go all the way to the heterotic string before you can actually distribute charge along the string. But in 1969, no one knew that, and the only method for putting charge in string theory relied on the work of two physicists whose names were Chan and Paton. They used to write equations which described the charges on the end. These charges on the end were called Chan-Paton Factors. Now it doesn't have to just be electrical charge on the end. For example, now that we've studied the quantum world, we know that there are additional charges such as the weak nuclear charge or the strong nuclear charge. Those would also be at the end of one of these open strings.

But if you choose something other than electrical charge, then the particle that I was calling the photon must be replaced by the appropriate object. For example, if we put color charge at the end of the string, then instead of a photon, we must be talking about objects that have properties like the gluons, the things that keep the quarks bound to the interior. Two very interesting observations have been made in recent years. By the way, you might wonder—wait a minute; weren't we talking about string theory? Why have we returned to talk about protons? Because the topic of this lecture is the research that's going on most currently in the field, namely, string theory is being used to try to understand how protons and particles like mesons actually work. There's a mystery about them. Let's look at this mystery for a second.

You've seen this drawing before. We've used it to talk about the way that the strength of a particle's interaction varies with energy. If you direct your attention to the strong interaction—remember we use a color-coded scheme—so the red dots that you'll see on our diagram correspond to the strength of the strong coupling constant. There it is. You can see we have gravity on the bottom. We have electromagnetism on top of that. We have the weak force, weak nuclear force, and the strong nuclear force. The strong nuclear force is the strongest. Now suppose we imagine looking at the particles that constitute a proton or a meson at much, much smaller energies. Then we go backward in the scale. We don't move to the right. We move to the left. If you look at the plot for the red dots, you'll notice something very interesting. They are shooting upward. What that means is that as you probe an object like the proton at smaller and smaller energies, the force which couples the quarks together must become stronger and stronger.

In fact, if you look at our three coupling constants, you'll realize the one for the strong interaction is the only one that's shooting up off the top of the graph. Of course, that's what we need because the quarks must be permanently bound to the interior of a particle like the proton, or one of the mesons or other forms of nuclear matter. There's a problem with this graph. You see, no one actually knows how to calculate that coupling constant that shoots off the edge. We think that's the right answer. We have lots of experimental evidence that that's the right answer. We know that as you probe to higher energies, that coupling constant decreases. In fact, a recent Nobel Prize was awarded for the work which was about the decreasing of the coupling constant. This problem of trying to understand why quarks can never get out of the interior of hadron matter reduces to understanding why does this coupling constant for the strong forces shoot off to infinity as we probe to smaller and smaller energies?

Physicists are very uncomfortable when we cannot explain something. So in recent years, some new ideas have come about trying to understand that strange behavior. Interestingly enough, these ideas have actually come from string theory. Remember string theory that describes the entire universe? In recent times, because of these two new developments, it has returned to its point of origin, which somehow completes a very nice circle, historically. Part of this work comes to us from three physicists in 1986, Kawai, Lewellen, and Tye. What they showed is something very interesting. You'll remember that closed strings always have a mode of vibration that looks like the graviton. You'll perhaps also remember that, at an earlier lecture, I showed you that a closed string, which is our standing wave mode, can always, in some sense, be decomposed into left movers and right movers. Perhaps you'll want to review that lecture to go over that point again, but it's actually the case.

This 1986 observation was that well, wait a minute. The left movers and right movers are what you need to describe open strings. But open strings describe things like the photon or the gluons. Therefore, there must be some very deep hidden mathematical relation between a theory of gravity—that's what closed strings provide—and the theory where the movers are either left movers or right movers, but those are equivalent to open strings. So there's something, some serious connection, between these two things that no one had ever thought about. If we review for a moment what we mean by right movers and left movers—let me put some images up for you. Remember, we use our red dots to tag the way the string is rotating. Here we see a typical picture of a right mover, or we can look at the picture of the left mover. It's rather similar. The only difference is that the rotation occurs in the opposite sense, and so we see the red dots moving in the opposite sense.

Each of those separately has something to do with photons or gluons or other force carriers that we see in the Standard Model, but not the graviton. There's this mysterious connection. In fact, this mysterious connection is so powerful and so useful that it's now called the KLT observation, or the KLT relation. KLT stands for Kawai, Lewellen and Tye, the physicists who first noted this mysterious connection between gravity and gauge particles like the photon; and in particular, for the gluon. This more accurate understanding that we're looking for—remember, we want to find an explanation for why is it that the quarks are the only particles whose force carrier has a coupling constant which runs off into infinity as we look at smaller and smaller scales.

The conventional picture, which physicists have been discussing for about 20 years, goes as follows. If we think about the hadron as a bubble—we've seen pictures of it—imagine that you could grab that bubble and start to pull on it with the quarks inside. In particular, as you pull the parts of the bubble to greater and greater distances, then the bubble has to deform. Remember that when we write Feynman diagrams, the forces between the objects are always carried back and forth by messenger particles; these things we call gauge particles. For the strong force, these are the gluons. As you pull the object apart, you imagine, if you had some kind of super vision, we could actually see the gluons going back and forth between the two quarks.

Something is very interesting about gluons, which distinguishes them from photons. If I have two electrical charges, the photons are essentially allowed to fly around any path to go from one positive electrical charge to a negative electrical charge. They can take the direct route, or they can go 100 billion miles away and come back. For gluons, this is not allowed. The gluons have to bunch tightly inside of the bubble. So effectively, because of that, when you pack gluons together, they start to increase their energy. Einstein taught us that energy is the same thing as mass. So when you pull the quarks farther apart, the gluons increase their energy until finally, they have enough energy to convert that energy from the energy of a gluon, or masses of gluons, to become the energy in the form of new quarks.

This is very interesting because, if you have new quarks, then the bubble is allowed to split. Instead of having one meson, you wind up with two—one with a quark and anti-quark moves off to the right; one with a quark and anti-quark moves off to the left. Everything that I've told you is a quantitative view, namely, no one has ever done a calculation to prove that all of that works. We think that's how it works, but we don't know for sure. One of the things that this Kuwai, Lewellen and Tye relation has done for the physics community is to give it an alternative way to study this process. However, even more interesting things have been occurring in recent times.

In 1997, something totally new happened, and this totally new idea goes by the name AdS/CFT. Those are letters; they stand for something. The ADS part stands for anti-de Sitter space. We'll come back and explain what that means. The CFT part stands for conformal field theory.

Let's start with the AdS part. In the 1930s, a Dutch astronomer whose last name was de Sitter, was the first person to argue that we live in a universe where the curvature tensor—and I hope you remember the curvature tensor—is this thing that actually describes gravity. De Sitter's assertion was, in our universe, the curvature tensor is non-zero and positive. De Sitter actually had discussions with Einstein about this point because, after all, Einstein had invented general relativity. That's what we mean by AdS. On the other hand there's CFT. Let's talk about that for a moment. If we lived in a universe where there were only particles like gluons, then the only sources of energy that would be possible to describe must be the energies associated with gluons.

However, if you write down the mathematics of a theory that is only describing gluons, what you find is a very curious property. Imagine taking such a world and studying it with a microscope. What you would find is no matter how you look at the system with a microscope—whether you set your microscope to power 10, power 20, power 100, power 1,000—the physics would always look the same. Physicists have a name for this kind of a system; it's called a conformal system. The C in CFT is conformal. The field in CFT is actually the field of the gluons, and then, of course, T is theory. So AdS/CFT tells us something—that if you are studying a theory where there are only gluons that have this funny property, it is related in some mysterious way to an anti-de Sitter, geometry, geometry where the curvature is negative.

This amazing discovery was made by a physicist named Juan Maldacena from Argentina. He's actually a student of Edward Witten, who is one of the leaders of the string revolution. The discovery marked the beginning of what's called the "second string revolution." The first string revolution had occurred after Green and Schwarz discovered the magical 496. Why are these discoveries so important? In order for the explanation for the shooting up of the coupling constants to be derived from mathematics, we need to find some new mathematics. We have, for over 30 years, not been able to derive this result. Some new mathematics and some new ideas have appeared in the idea and in the form of objects that are called branes. Let's take a visit to the world of branes.

Here we have a transparency. If you look in the middle, you see a little dot. I want you to think about that dot, for the moment, as a charged particle. For example, that might be an electron. If it were a real electron, then of course, it would be free to move about. I'm going to give a new definition at this point. Instead of calling it a dot,

I'm going to call it a zero-brane. If that's a zero-brane, what's a one-brane? Now we have a line. In fact, that could be one of our strings. We're going to again introduce a new word. Whenever we have something that has one dimension associated with it, we're going to call it a one-brane. Why stop at one? Well, in fact, there's no need to. We can go to two dimensions. Let's take our string and roll it out. Now we have something that has a two-dimensional plane. We would call that a two-brane.

In our world, of course, the only other object one could write would be a volume, and that would be a three-brane. But in the world of the mathematics of a string, nothing stops you from doing this process all the way up to ten because, in string theory, the mathematics looks as if there are ten dimensions. One of the interesting things about this process, however, is that we started with the electrical charge, which was a single zero-brane. It was allowed to carry charge. Since we know how to control the mathematics, we can let these other objects carry charge. By the way, the collection of all possible branes is called Dp-branes, where p can be arbitrary, basically. We're going to talk about Dp-branes. If the electrical particles like the electrons couple to the photon to transmit forces, can these Dp-branes carry charge? And if they do, are there force carriers that can couple to them? The answer mathematically has been known for a long time; it's yes.

So these other objects also can give rise to electrical forces. This is very important because, if you look at the mathematics of these other objects, these Dp-branes is that strings end upon them. Dirichlet was a mathematician whose work is important to understand modern string theory, even though his work occurred almost 200 years beforehand. The important point about these Dp-branes is that strings end upon them. Remember we first talked about charges being at the end of the strings, but in fact charged Dp-branes can be at the end of strings. It was the physicist Joe Polchinski in 1990 who first focused the string theory community on studying these objects, even though another physicist by the name of Warren Siegel had actually first written his thesis about them in 1977. What's so neat about these Dp-branes? By using Dp-branes, you can actually prove that this weird connection between the theory where there are only gluons and some connection to a theory with only gravity actually exists. In fact, people have used the notion of Dp-branes to derive an equation which connects gravity to a theory that only involves the gluons.

There's a lot of excitement, in fact, in the physics community because, for the first time, instead of arguing in terms of pictures or words about why the coupling constant is shooting upward for the strong forces—for the first time, physicists are now being able to calculate this process of the coupling constant going up. In fact, Edward Witten has made an even more startling suggestion about how string theory may contribute to our understanding of the strong force. This last suggestion of Witten's is curious in many, many ways. Let's get it up on the screen. It's going to involve something that we already know. You'll remember the notion of anti-commuting numbers. In fact, let's put it back up so we can see it. Ordinary numbers, when you multiply them, don't care about the order in which you multiply them. Four times three is the same thing as three times four.

In the 1800s, the mathematician Grassmann decided to study numbers that had a different property; namely, when you multiply them, you get a minus sign. In talking about string theory, this second kind of object where the minus sign occurs leads to a description of string theory, which is called the Neveu-Schwarz-Ramond Theory, the NSR Theory. In the NSR Theory, you have the strands which describe the string, but you also have different mathematical strands that have this kind of multiplication law attached to them. It turns out that you can, in fact, do something very interesting; namely, for every strand of the string that's there, introduce a mirror copy of it that possesses the wrong multiplication law. When you do that, you can produce—you can say well, I will do one mirror copy for every strand—that's the $N=1$ Neveu-Schwarz-Ramond string—or you can introduce two copies. If you introduce two copies with the anti-commuting multiplication, you get what's called the $N=2$ Neveu-Schwarz-Ramond string. From this object emerges an even more interesting beast.

I hope you remember spinors. An electron is a spinor. It's this notion that there are objects in nature that you have to walk around twice before you find out that they're pointing at you again. We talked about this a long time ago, but it's relevant here. In these so-called $N=2$ theories, spinors also exist, but they have another name entirely. They're called twistors. Why did we make this excursion into twistors, of all things? Twistors have, first of all, been around for a very long time. In fact, they were first proposed by the physicist

Roger Penrose in the 1960s. But usually, people have only tried to use twistors to understand various aspects of gravity. You perhaps won't be surprised because of our AdS/CFT connection, where we learned that gravity is connected with gauge theory and gluons. You might begin to suggest that gee, maybe twistors are useful. In fact, that's precisely what Edward Witten has done in an impressive paper quite recently.

There's one other very mysterious thing that I have to tell you about, this so-called N=2 string. Remember, the basic idea is that for every strand of the string—it has an X direction, a Y direction, a Z direction and six friends—you introduce two copies that are anti-commuting quantities. If you do that and ask what the Einstein hypotenuse is for the notes produced by such an object, you discover something quite remarkable. The notes obey the Einstein hypotenuse, but only if there are two time dimensions. Even though you may have never heard about the discussion that says that strings can produce theories where there are two times, such objects are well known. It turns out that twistors are exactly connected to them.

In this way, a lot of rather esoteric mathematics is starting to be marshaled for the purpose of understanding not some kind of particle we've never seen—remember, we've never seen a superparticle. We've never seen a squark or a slepton or a gaugino. We've never seen those superpartners that we've talked about, but we have seen nuclear matter. We know protons, we know neutrons, and we know mesons and the rest of the nuclear zoo. So string theory, while many people like to say it's not something that can be tested, there is this new horizon on string theory where it is, in fact, the only piece of mathematics that we have acquired that has allowed us to calculate a physically interesting property; namely, the running of the coupling constant. The next time someone tells you that string theory is not testable, remind them of the AdS/CFT connection and how, in fact, physicists right now are starting to test string theory in a new and exciting way.

Have we finished yet, you might ask. Well, not quite. Even though this AdS/CFT notion has started to yield something that looks like real physics, it's not quite there yet. What's the problem, you might ask. AdS/CFT is an interesting and exciting new idea. The only way we know how to actually test this idea is to work in what's called a type IIB string. I know you're holding your head by now, saying wait a minute, what is that? In the next lecture, we're going to actually talk about the IIB string. In fact, we're going to talk about all of the strings. There are going to be five of them in toto, and I'm going to explain what they are. This particular one, the IIB string, was initially thought to describe only gravity. But the AdS/CFT connection tells us remember that gravity, in some mysterious way, can be connected to the world of gluons. So the actual use of the AdS/CFT theorem to date has only occurred in the context of the IIB string. When you look at the IIB string, there's only one problem.

As I've described supersymmetry for you, it's the notion that for all the ordinary matter that you know, there's a mirror partner. You could say well, wait a minute, why only one partner? Why not many? In fact, there can be, at least in the realm of mathematics. Instead of having just one partner for the electron, there could be two partners. We call that $N=2$ supersymmetry. There could be four partners; that would be called $N=4$ supersymmetry. In the same way, for a gluon, there might be not just one gluino partner; there might be two or perhaps four. When you look at the actual mathematics that comes from the AdS/CFT construction, you find out that for every gluon, there are actually four gluinos. That means it's $N=4$ supersymmetry. Well, it's a hard task to accept that for all the matter that we have seen in the world, there's a hidden superpartner. But it's even a greater leap to say that there must be four of these things.

So although mathematically, $N=4$ supersymmetry is very well-defined and is the actual mathematical tool by which we have been exploring the AdS/CFT connection, in point of fact, we're not quite there in looking at our world. One thing we're going to have to do, at least as a beginning step, is to learn how to get rid of the other three superpartners that we don't believe should be present. Nonetheless, this leads to the most precise calculations we have to date of the running coupling constant and QCD, quantum chromodynamics, and it is not string theory talking about some mysterious other particle. It's string theory talking about our world.

Lecture Twenty-Two
If You Knew SUSY

Scope:

The expression *SUSY* means supersymmetric or supersymmetry. This lecture will lead to a rather unusual aspect of SUSY—*superspace*. Ordinary particles are free to move into superspace, but when they do, they become superpartners. The idea of superspace combines with the notion of strings to explain the existence of five types of superstrings.

Outline

I. What if there were more bizarre kinds of hidden dimensions?

   A. In 1974, Salam and Strathdee invented a new kind of space that could accommodate our dimensions as one of its parts as well as these new bizarre dimensions. This construction is called *superspace*.

   B. In our world, if we want to buy carpet for a room, we have to measure and multiply the dimensions of the room, but the order in which we measure the dimensions of the room doesn't matter. With algebra, it is possible to write an expression for this: $x_1 \times x_2 = x_2 \times x_1$. With this expression, we see that lengths are commuting numbers.

   C. The work of Salam and Strathdee argues that, in order to describe physics, we may sometimes require numbers that correspond to the lengths that are commuting numbers, but they also investigated the possibility that some of the extra dimensions might not be numbers at all in the typical way numbers are used and envisioned.

II. What prompted Salam and Strathdee to look into this idea was something that was already happening in string theory.

   A. Salam and Strathdee were looking for a simple way to understand how bosons could be traded for fermions.

   B. They hit upon an idea that involves extra dimensions. They argued that extra dimensions existed, but they used Grassmann numbers instead of ordinary numbers for these extra dimensions.

C. Salam and Strathdee introduced four of these extra dimensions, called θ_1 through θ_4. If any two of these are multiplied, the result obeys Grassmann's anti-commuting rule. By these definitions, then, $\theta_1 \times \theta_2 = -\theta_2 \times \theta_1$. More remarkably, the square of any one of these numbers vanishes even though none of them is, itself, 0!

D. Why did Salam and Strathdee introduce four of these extra dimensions?
   1. In the discussion about the electron, we stated that it has 4 degrees of freedom. It can spin in either a right-handed sense or a left-handed sense, and its antiparticle can also spin in either a right-handed or a left-handed sense.
   2. One might be tempted to think that the antiparticle has nothing to do with the electron. But the mathematics that describes an electron and its antiparticle (the *positron*) can be mapped one onto the other if the antiparticle is interpreted as moving backward in time. This remarkable result was found by Richard Feynman.
   3. This means antiparticles are not distinct from the particles they mirror; they are the same particles moving backward in time.
   4. In that case, it makes sense to say that an electron spinning in either a right-handed or a left-handed sense is still an electron and a positron spinning in either a right-handed or a left-handed sense is also still an electron. Thus, an electron has 4 degrees of freedom.

E. Salam and Strathdee were not the first people to introduce anti-commuting numbers into physics. This honor belonged to a Russian physicist named Felix Berezin, who was more interested in practical problems in physics. Strangely, he found that introducing anti-commuting numbers actually simplified his calculations. Berezin is the father of what is now called *superanalysis*.

III. What is superspace? Superspace is a construct that has the coordinates of our world (duration, breadth, length, and thickness), along with four anti-commuting extra directions. This construct is called *N = 1 superspace*.

A. More complicated versions of superspace are possible. We could imagine a mathematical construction that includes the coordinates of our world and adds two copies of these anti-commuting directions. This construct is called $N = 2$ *superspace*. We might also imagine repeating this process to arrive at what is called *N-extended supersymmetry*.

B. One of the more remarkable things about this construction is that the equations of ordinary physics can be rewritten over superspace.

C. However, there is an upper limit to superspace.
   1. Many lectures ago, we introduced the fact that the force of gravity is somehow connected with the curvature of space and time.
   2. When this idea of Einstein's is applied to superspace, a limit is found. The highest number of copies of the anti-commuting coordinates that can be introduced is eight!

IV. Ordinary particles moving in ordinary space and time don't change their identities. In the mathematics of superspace, however, if an ordinary electron moves into one of these extra directions, the electron becomes one of the selectrons!

   A. This gives us a simple understanding of why superpartners appear, and it is similar to Feynman's earlier observation that one way to understand the presence of antiparticles is to think of particles moving backward in time. In the notion of superspace, when an ordinary particle moves into one of the superdirections, it becomes a superpartner.

   B. This means that supersymmetry is just a statement that ordinary particles are free to move in any direction in superspace. In fact, in the mathematics of superspace, it is straightforward to show that the superpartners obey exactly the equations of our world.

V. There is one funny aspect of moving around in superspace.

   A. Imagine some children playing with a ball on a soccer field. One child starts in one corner of the field and kicks the ball; the ball rolls along the boundary of the soccer field until it reaches the next corner and a child there stops it. That child kicks the ball in the orthogonal direction, and it then travels along the boundary until it is stopped at the next corner. The children continue this game until the ball has traveled around the perimeter of the soccer field. When the ball gets back to the first child who kicked it, it's still a ball.

B. Imagine now that these children have access to superspace. The first child kicks the ball in one of the directions of superspace, but when it reaches the second child, it's no longer a ball; it's a superball. When the second child kicks the superball to the third child, it changes back to an ordinary ball. Ultimately, when the ball reaches the first child again, it has changed four times, from ordinary ball to superball, back to ordinary ball, back to superball, and finally, back to ordinary ball.

C. One might think that the first child wouldn't notice any difference in kicking the ball in a complete circuit in ordinary space directions vs. superspace. However, the equations of superspace indicate that when the ball returns to the first child, it is hovering above the ground!

D. In this story, superspace is the soccer field, but the area above superspace is ordinary space. In moving around in superspace, whenever you try to return to the starting point after moving in superspace, the ball returns back into ordinary space in a different position. Mathematicians call this property *torsion*.

VI. Does Einstein's idea about curvature apply to superspace? Remember, in ordinary space, the notion of curvature of space and time leads to a description of gravity. If superspace is so much like our space, can it support curvature?

A. This question was answered in 1997. Wess and Zumino, and later Deser, found it is possible to write a curved supergeometry—a construct that incorporates Einstein's notion of curvature into superspace.

B. In ordinary space, incorporating curvature leads to gravity. In superspace, incorporating curvature leads to supergravity. Recall that the existence of the gravitino was important to explain why the superpartners are so much heavier than ordinary matter. Thus, application of the ordinary laws of physics to the strange mathematics of superspace generates a consistent view.

- C. One can also apply the concept of superspace to Maxwell's theory, and the result is photons and photinos. In fact, every single equation of fundamental physics discovered for the past 150 years can be described in the mathematics of superspace and results in the superpartners.
- D. In Einstein's quest to explain gravity, he had to learn new mathematics to describe curvature. This same mathematics more or less exists for superspace. But the discovery of Einstein's equations in their most fundamental form in superspace presented a problem for a short while. Warren Siegel and I worked together at Harvard in 1977 to find the superspace analogs to Einstein's equations.

VII. The idea of superspace explains the fact that there are five strings.
- A. An open string added together with 16 of these Grassmann coordinates that support vibrations moving in a left-handed sense is an *open type-I superstring*. Why do we add 16 Grassmann coordinates? The answer is that a spinor in 10 dimensions must have 16 components, not 4.
- B. A closed string added to 16 Grassmann coordinates that support both left-moving and right-moving vibrations is a *closed type-I superstring*.
- C. A closed superstring added to 32 Grassmann coordinates results in a *closed type-IIA superstring*. In this superstring, the 10-dimensional spinors are also both left-handed and right-handed and support left-moving and right-moving vibrations.
- D. A closed superstring added to 32 Grassmann coordinates results in a *closed type-IIB superstring*. In this superstring, the 10-dimensional spinors are all left-handed and support left-moving and right-moving vibrations.
- E. Finally, the closed string added to 16 Grassmann coordinates that move only in a left-handed sense plus the angles associated with gauge theories leads to the heterotic string.
- F. No other consistent ways exist to play this game. Counting the number of formulations that come out of this theory leads precisely to five.
- G. Let's review these five superstrings again:

1. Open superstring + 16 left-spinning anti-commuting numbers (that support only left-moving vibrations) = open type-I superstring.
2. Closed superstring + 16 left-spinning anti-commuting variables (that support both left- and right-moving vibrations) = closed type- I superstring.
3. Closed superstring + 16 left-spinning anti-commuting variables (that support both left-moving and right-moving vibrations) + 16 right-spinning anti-commuting variables (that support both left-moving and right-moving vibrations) = closed type-IIA superstring.
4. Closed superstring + 32 left-spinning anti-commuting variables (that support both left-moving and right-moving vibrations) = closed type-IIB superstring.
5. Closed superstring + 16 left-spinning anti-commuting variables (that support only left-moving vibrations) + a number of right-moving vibrations that describe either SO(32) or $E_8 \times E_8$ = heterotic string.

VIII. In the last lecture, we considered the fact that string theory is now being used to explain how quarks and gluons are held in the interior of nuclear matter. The ending point for the last lecture was AdS-CFT, also known as the *Maldecena conjecture*.

   A. Your lecturer's own work was undertaken in the context of a IIB string. Thus, this theory requires a IIB string and an anti-de Sitter geometry (i.e., a space with negative curvature). The output of this theory is not a theory of gravity, but a theory that contains the gauge fields that occur in the standard model. This shows how current string theory is attempting to make connections to the real world.

   B. In 1986, working with two of my students, Roger Brooks and Floyd Muhammad, we showed that the dimensions associated with forces in our world exactly follow the patterns that are appropriate for string theory. This is the work that permits the exact appearance of the Kemmer angles in string theory

   C. For anyone who has endeavored to follow this lecture, it can be said that *you* know SUSY; it is a strange construction that has hidden directions, but the mathematics of superspace indicates that for every particle seen, there is one superpartner…if our universe is supersymmetric.

Readings:

Greene, *The Elegant Universe* and *The Fabric of the Cosmos: Space, Time and the Texture of Reality*.

Kaku, *Hyperspace: A Scientific Odyssey through Parallel Universes, Time Warps, and the 10th Dimension*.

Kane, *Supersymmetry: Unraveling the Ultimate Laws of Nature*.

Questions to Consider:

1. Explain how the idea of superspace leads to the superpartners.
2. Describe the five types of superstrings.

Lecture Twenty-Two—Transcript
If You Knew SUSY

If you were to pick up a physics journal anytime in the last 20 years—and I don't mean one such as *Physics Today*, but the hardcore journals that the physicists actually use to do their research—you would likely come across the word SUSY, S-U-S-Y, usually in capital letters. This means supersymmetric. In this lecture, which is titled "If You Knew SUSY," we're going to try to understand a rather unusual aspect associated with this. In many of our lectures, we've talked about the possibility of hidden dimensions. Maybe our world has more directions than we are used to thinking about and that we experience in everyday life. In some ways, even though a bit odd when first one hears such an expression or such an idea, these extra dimensions, these hidden dimensions, if they're there, are pretty conventional. They're kind of like the space that we already know. It's just more of it. What if there are even more bizarre kinds of hidden dimensions? What else could there be?

In 1974, a physicist from Pakistan, Abdus Salam, working together with John Strathdee of New Zealand, proposed a really radical answer to this question, what other kinds of hidden dimensions might there be? They, in fact, had to invent a new kind of space to accommodate our kinds of dimensions; as well to accommodate these extra bizarre dimensions that are even more bizarre than the hidden ones we've discussed so far. They called this construction superspace. In our world, we can make measurements—again, I'll go back to an example I've used several times before, putting carpet on the floor—if you want to go to the store and buy carpet to cover a floor, you have to tell the salesperson how much carpet you want. If you have a room which is five feet by eight feet, then you multiply those two numbers together, and you need 40 square feet.

The other thing that's interesting about this process is which measurement do you take first? Do you have to measure the length that's eight feet long, or do you have to measure the length that's five feet long? It doesn't matter. You can measure the eight, measure the five and do the multiplication. Five times eight is eight times five, so it doesn't matter which one you measure first. But now let's use that algebra again. We can write an equation, which says eight times five is equal to five times eight, and everyone will agree with that. But if we use algebraic symbols for these things, where the number eight, we'll call X_1. The number five, we'll call X_2. You get an equation which says X_1 times X_2 is equal to X_2 times X_1.

But wait a minute. We've seen this kind of thing before. If we subtract the X_2 times X_1, then we see that, wait a minute, lengths are commuting. They're like these commuting numbers that we keep talking about. So the obvious question that this raises is is it possible to have some kind of a universe where lengths do not behave as ordinary numbers that we're familiar with, but instead behave as anti-commuting quantities, much as electrons are anti-commuting? This is a very bizarre idea. In fact, no one had ever thought of such a thing before 1974. So the work of Salam and Strathdee is exactly this idea, that in order to describe physics, not only do we need numbers that correspond to the lengths that we know that are commuting objects, but we also have to investigate the possibility that some of these extra dimensions that are popular in string theory might not be numbers at all in the typical way that we think of numbers.

The reason why Salam and Strathdee did have such a thought had to do with something that was going on in string theory already. Remember, string theory gets its start at the end of the 1960s. In 1974, string theory, although not very popular and also not very well studied, had a few people that were investigating it. Among the results that were found from these investigations was this strange thing that you're allowed to trade bosons for fermions. You can trade something that's like an electron for something that has properties like the photon, a very different and new and exciting idea at the time.

So Salam and Strathdee were looking for a simple way to understand how this was possible at all, and they hit upon an idea. The idea is quite simple. It involves extra dimensions, so we have to have one variable that stands for time, one for what we physicists like to call the X direction, another variable for the Y direction, a third for the Z. So you have to have four variables that talk about the position in space as we understand it. Then they went beyond that. They said there are extra dimensions, but we're not going to use numbers for these extra dimensions. Instead, we're going to go back and use Grassmann numbers as first discussed by the mathematician Grassmann in the 1800s.

So they didn't just introduce one of these objects, they introduced four. We can write these objects, again, using algebra. Remember, algebra for us is a picture, a tool for thinking about ideas. We can introduce four of these objects. We'll call them θ_1, θ_2, θ_3, and θ_4. If we were to multiply any two of them, we get an anti-commuting object. θ_1 times θ_2 is, by definition, $-\theta_2$ times θ_1. Or if we multiply θ_1 times θ_3, that's equal to $-\theta_3$ times θ_1. More remarkably, the square of any one of these objects is vanishing. So these certainly are not numbers that we're used to in any usual description of nature, but they're going to be useful in talking about supersymmetry. You should have a couple of questions; why four? That one's pretty simple to answer. You'll remember, when we were talking about the electron, I said that the electron kind of has four degrees of freedom. What did I mean by that?

We can talk about an electron—that it can spin in a right-handed sense or a left-handed sense; it's still an electron. Interestingly enough, the anti-particle to the electron can either spin in a right-handed or left-handed sense; they're four degrees of freedom. But you might be tempted to say that the anti-particle has nothing to do with the electron. Richard Feynman actually settled that a long time ago. If you look at the mathematics which describes an electron and then look at the mathematics that describes its anti-particle, also called the positron, those two pieces of mathematics can be mapped, one into the other, if you interpret the anti-particle as being the particle moving backward in time. Again, a very strange notion, but mathematically precisely defined.

That means that anti-particles are not, in some sense, distinct from the particles which they mirror. They are in some sense those particles moving backward in time. If that's the case, then it makes sense to say that the electron, spinning in either a right-handed or left-handed sense, is still an electron, and its anti-particle, the positron, spinning in either a right-handed or left-handed sense, is still an electron. So there are four different versions. That's why Salam and Strathdee introduced four of these numbers. Remember, another property of the electron, in addition to the fact that it obeys the Pauli Exclusion Principle, is that it's a spinor.

A spinor, well, it's—think of an arrow. It's this funny kind of arrow that we cannot actually build here in our universe, in the world of the everyday experience, but it's an arrow that you have to walk around either two times or four times or six or eight times to find out that it's pointing at you if it was before you started. But if you walk around it any odd number of times, it's actually pointed away from you. That was our basic working definition of a spinor. Since the electron is a spinor, and it takes four degrees of freedom to make up an electron, then Salam and Strathdee reasoned that if they wanted this extra dimension to be a spinor, you needed four of them. So that's why they introduced four.

They were not the first people to introduce anti-commuting members into physics. This honor belongs to a Russian physicist named Berezin. Berezin was concerned not so much about particles, but he wanted to know about reasonable properties such as how current flow through wires. How does heat flow? What is the tensile strength? When you take a wire, you're concerned about how to bend it, how strong it is. So these are real practical problems to be concerned about—how gases conduct heat, practical problems. But for some reasons and for some calculations, Berezin found that introducing the notion of these strange numbers actually simplified the calculations.

Berezin is truly the father of what's now called superanalysis. That means doing mathematics with these things that have this peculiar anti-commuting property. However, Salam and Strathdee went beyond Berezin because they wanted four of these. Berezin had never introduced more than one of these objects to simplify his calculations on the properties of ordinary matter. What's a superspace? Something that has the coordinates of our world—we can think of that as duration, length, breadth and thickness, or physicists would say something like time, X direction, Y direction and Z direction. They mean the same thing. In order to make it a superspace, we put four of these extra anti-commuting numbers as directions. That's what's called $N=1$ superspace.

More complicated versions of superspace are possible. For example, we could imagine taking a mathematical construct that includes the coordinates of our world—duration, length, breadth and thickness— and put two copies of these anti-commuting numbers. So we go from one to four in one copy, and then from five to eight for a second copy. Such an object would be called $N=2$ superspace. You could imagine doing this more and more times. You could have higher and higher copies, and that's called N-extended supersymmetry, where N is whatever value you decide to pick. One of the most remarkable things about this construction is that you can take the equations of ordinary physics and rewrite them over this other thing, this superspace.

The equations that you get look remarkably like our equations. In fact, not only remarkably like our equations—they are our equations, even though you started by saying these extra directions are there. There's an upper limit, however. This upper limit occurs in a very strange way. In our world, there's a force of gravity. Many lectures ago, we learned that Einstein's great genius teaches us that gravity is somehow connected with the curvature of space and time. So if you actually try to take this idea of Einstein and apply it to superspace, you run into a limit. This limit says that the most number of copies of the anti-commuting coordinates that you can introduce is eight.

Why is superspace interesting? Well, in our world and in ordinary space-time, particles can move about, and you can measure their properties. So if you believe that Salam and Strathdee constructed something that's interesting, one question you could ask yourself is what happens if an ordinary particle moves into one of these strange directions they've proposed? If you study the mathematics, it's amazing, namely, suppose I had an ordinary electron. If I have an ordinary electron, I could move it front and back; it's still an electron. I can move it left and right, still an electron. I can move it up and down, and it's still an electron. If I wait two seconds and look at it, it's still an electron. So for ordinary particles moving in ordinary space, they don't change their identity. But remember, Salam and Strathdee said there are these other directions. So what happens if we take an electron and move it into one of these other strange directions? Remember, we could only do this with mathematics. It's not that we can tell a story; we actually have to study equations to answer this question.

There's an answer that one can find from these equations. Remember in the world of supersymmetry, the electron has four superpartners; they're called selectrons. What happens in the mathematics of superspace is if you take an ordinary electron and move it into one of these extra funny directions, you find out that the electron changes its identity. In fact, it becomes one of the selectrons. So we, in fact, have a simple understanding of why the superpartners appear if you believe this superspace exists. Remember, a moment ago I described Feynman's observation that one way to understand the presence of anti-particles is to think of particles moving backwards in time. We now have a relatively similar observation. If we use the notion of superspace, when an ordinary particle moves into one of the super directions, it becomes a superpartner.

Now, of course, we can see why we had to have that balance that I talked about when we looked at the Standard Model. Remember, for every ordinary particle, there was a superpartner. If all that's happening is that the ordinary particles are moving into these strange directions, then for every ordinary particle, after it has moved into one of these strange directions, it would be a superpartner. That's the reason we find this nice balance. That means that the notion of supersymmetry, namely that we have this balance here, is just a statement that ordinary particles are free to move in any direction in superspace. That's a nice geometrical result. In fact, we can use it in order to rewrite all the equations with a sense of geometry. We have an intuition about how geometry works in our space. Now we can apply that intuition about geometry to this larger construct. When you do that, that's how we discover that the superpartners, in fact, obey the equations of our world, because we apply exactly the same geometric intuition.

There is one funny and very, very strange thing, however, about moving around in superspace. Let me tell you a story. Let's imagine I have a soccer field, and there is a ball on a soccer field, and there are some children who wish to play a game. One child decides that they're going to kick the ball in that direction, and they'll follow the boundary of the soccer field. So you start in one corner. One child kicks the ball. It rolls along the boundary of the soccer field until it gets to the other edge, and there's a child there that stops it. That child then kicks the ball along the orthogonal direction, the direction at right angles. And the ball travels around the soccer field, and a child stops it there at that end. That child then kicks it to another child, who then kicks it back to the first. This is something that's easily possible to imagine. When the ball gets back to the first child who kicked it, it's still a ball. Because after all, in our world, just by moving doesn't change the identity of something.

Let's imagine that these children have access to superspace. So the first child kicks the ball into one of these strange directions. When it gets to the second child, it's no longer a ball, it's a super ball. The second child kicks the ball at the direction that's right angles. When you do that, since it starts off as a super ball, by the time it gets to the child over there, it's an ordinary ball again. That child then kicks the ball along a third leg of the field. By the time that it gets to this third child, it's again a super ball. Now the fourth child kicks it back to the first, it turns back into a ball. So you would say well, the first child would therefore not notice any difference in moving around in ordinary space versus moving around in superspace.

There's only one problem with that. Remember, the only way we can answer these questions is by actually using equations. So we do the equations that tell that story, and what we find, quite remarkably, is that when the ball gets back to the original child who kicked it, it's up off the ground. It's not there because any of the children tried to kick it up. It has no choice. It must come back off the ground. What does it mean to be off the ground? Well, in my story, superspace was the soccer field. Up is where we live, up would be ordinary space. So what happens in superspace is a very strange process. You try to move around in superspace, and whenever you try to get back to the point where you started, you move back into ordinary space in a different position.

Mathematicians have actually been studying structures such as this, not in superspace, but in spaces where we have ordinary coordinates. Whenever this kind of story happens, mathematicians have a very special name for that kind of a space. They say it has a property that is called torsion. Superspace has torsion, but it's not the only space that has it. Superspace turns out to be a fairly gentle excursion away from the mathematics of our world, one reason why it's very simple to explore the possibility of physics in such a construct. Now we can ask a really important question. Does Einstein's idea about curvature apply to superspace? Remember, in ordinary space, it's the notion of curvature of space and time that leads to a description of gravity. So if this superspace is so much like our space, can it support curvature?

In 1977, this question was answered, first by Julius Wess and Bruno Zumino, and later by Stanley Deser. They found that you could write a curved supergeometry. This is a construct that uses superspace and also incorporates Einstein's notion of curvature. In ordinary space, when you incorporate curvature, you get led to gravity. In superspace, when you incorporate curvature, you get lead to supergravity, but in fact, we've already met supergravity. In our lecture where we explained how it was that the superpartners could have been missed in experiments, we said they had to grow a mass.

In order for them to grow a mass, we actually had to introduce a partner to the graviton that ate up what we called the Goldstino. Remember, the Goldstino was something that actually caused the mass. But the Goldstino itself is also not visible, so something has to more or less devour it. This object that devours the Goldstino is called the gravitino. It is the actual superpartner to the graviton. It's actually a consistent view that we have. We apply ordinary laws of physics to this very strange mathematical construction, and out pops the superpartner to the graviton.

The other thing about this sort of idea is you can say well, if you can do this for gravity, can you do it for Maxwell theory? The answer is yes; that's where we get photons and photinos. In fact, every single equation of physics that we have discovered for about 150 years, we can apply in superspace, and all that comes out is the superpartner. In Einstein's quest to explain gravity, he had to learn new mathematics to describe curvature. This same mathematics, more or less, exists for superspace. But the discovery of Einstein's equations in their most fundamental form was, in fact, a problem that persisted for a short while. I was, in fact, very fortunate to be working with my friend Warren Siegel at Harvard in 1977, and we found the superspace analog of Einstein's equations.

What does this all have to do with strings? Well, now I can tell you why there are so many strings. Why is it? Well, remember we talked about the open string. Suppose we take an open string and add to it 16 of these Grassmann variables. Why 16? Because a spinor in ten dimensions has to have 16 components, not four, so you add 16 of these Grassmann coordinates. Also suppose that these coordinates support vibrations that move in a left-handed sense. If you can do that, you have what's called a type I superstring. Suppose that you take a closed string and again add 16 Grassmann coordinates, but you allow their motions to be both left moving and right moving. If you do that, you get the closed type I superstring. If you take a closed superstring and add not 16, but 32 of these Grassmann coordinates, where their vibrations are both left and right, you have what's called a closed type IIA superstring. If you do this in a slightly different way, again adding 32 of these extra coordinates, you have the closed type IIB string.

The IIB string is actually different from the IIA string because the spinors are all left handed in the latter, whereas in the first, some are left and others are right. Then finally, take the closed string, add 16 of these Grassmann coordinates, make sure that they only move in a left-handed sense, and then add in the angles that we saw associated with gauge theories, which, in fact, in a previous lecture, when we discussed the dimensions. They are not the same angles. That leads to the heterotic string. It turns out that there are no other consistent ways to play this game. So when you actually count the number of theories that come out of this formulism, the number turns out to be precisely five.

Around 1995, if you had read the literature on string theory, you would have found people talking about five different superstrings. To review these five superstrings, we can take an open superstring, add to it 16 left-spinning anti-commuting numbers that are also only left-moving vibrations; that gives us the open type-I superstring. Or we can take the closed string, add 16 left-spinning anti-commuting variables, however, that support both left and right vibrations; that gives us the closed type-I superstring. The next alternative is to start with the closed string, add to that 16 left-spinning anti-commuting variables, which support vibrations that are both left-moving and right-moving, and add to that 16 more right-spinning anti-commuting spinning variables, which both support left-moving and right-moving vibrations; that gives us the closed type-IIA superstring. This is followed by the closed type-IIB superstring. We start with a closed string, we add 32 left-spinning anti-commuting variables, which nonetheless support both left and right rotations of the string; this is the closed type-IIB superstring. Finally, we can start with a closed string, add 16 left-spinning anti-commuting coordinates that only support left-rotating modes of vibration, and then add to that right-moving modes of vibrations that describe either $SO(32)$ or E_8; this gives us the heterotic string.

In our last lecture, we talked about the fact that currently, string theory is not all just trying to explain the world; it's trying to explain how quarks and gluons are held in the interior of nuclear matter. The starting point for this construction was the AdS/CFT conjecture; also, by the way, known as the Maldacena conjecture, named after the young physicist who first had the idea. One of the strings that I just introduced you to, at some level, was a IIB string. The work of Maldacena was precisely in the context of the IIB string. So what Maldacena actually says is you have to start with the IIB string. You have to work on a geometry which is anti-de Sitter, so it's a space that is curved. The output of that is not a theory with gravity, but a theory that contains the gauge fields—things such as gluons, things such as photons, things such as the intermediate vector bosons—that occur in the Standard Model.

Now we, in fact, know something more about how current string theory is attempting to make connections with the real world. It's not all about gravity. It can be about quarks. In the heterotic string case, I made a small contribution myself because the Kemmer angle work, that is, the work where you introduce the dimensions. You'll remember dimensions mean many things for us now. In particular, it means if you have a force carrier, it's the change in the force carrier that causes a force. You can have two different force carriers whose change is the same, but they're different. So you need some way to relate them. We also call that way a dimension.

In 1986, working with two of my students, Roger Brooks and Fuad Muhammad, we were able to show that these dimensions that are associated with forces in our world exactly follow the patterns that are appropriate for string theory. So now we know what SUSY is. The title of the lecture was "If You Knew SUSY." Well, we do know SUSY. It's a strange construction. Again, it says our world has hidden directions. In the superspace construction, however, the hidden directions are not numbers; they're Grassmann numbers, objects whose square is equal to zero. They're much gentler than ordinary numbers, so doing physics over superspace actually produces the kind of physics that we know, but with a difference. It says for every particle we have seen, there's one superpartner.

Lecture Twenty-Three
Can I Have That Extra Dimension in the Window?

Scope:

The last lecture offered an explanation of the assertion that there are five different strings. There's a problem with this assertion: Strings supposedly describe everything. If so, how can there be five different "everythings"? An answer to this question may be found in a construct that is not a string theory, *11-dimensional supergravity*. A look beyond 11-dimensional supergravity suggests it is part of a larger and even more mysterious construct, *M-theory*.

Outline

I. The starting point of this discussion is a review of the idea of a *mode*.

   A. An animation shows a closed string vibrating in such a way that three lobes seem to appear and disappear as the string vibrates. We can also find strings in which more or fewer lobes seem to appear and disappear. The number of lobes seen on the string is what we mean by a *mode of vibration*.

   B. Every time the string vibrates in a different way, from our perspective, we see a different particle. Rather than talking about the entire string, then, it is possible to consider only its simplest mode of oscillation—the one in which only two lobes would appear in our picture.

   C. If we restrict ourselves to this simplest way of vibrating, a full consideration of the entire string is no longer the focus; only a part of the string is the subject of this part of the lecture. This idea can be applied to the five strings that were known in 1994: the open type-I superstring, closed type-I superstring, closed type-IIA superstring, closed type-IIB superstring, and heterotic string.

   1. Truncation of the open string, and considering only its lowest mode of vibration, leads to 10-dimensional supersymmetric Yang-Mills theory.

   2. Truncation of the closed type-I string, and considering only its lowest mode of vibration, leads to 10-

dimensional $N = 1$ supergravity + super-Yang-Mills theory.
3. Truncation of the closed type-IIA string, and considering only its lowest modes of vibration, leads to 10-dimensional $N = 2A$ supergravity.
4. Truncation of the closed type-IIB string, and considering only its lowest mode of vibration, leads to 10-dimensional $N = 2B$ supergravity.
5. Finally, truncation of the heterotic string, and considering only its lowest mode of vibration, leads to 10-dimensional $N = 1$ supergravity + super-Yang-Mills theory, but there are some slight differences here from the truncation of the closed type-I string.
6. In all these truncations, the world of strings has been left behind, and the equations investigated are more similar to the equations that describe the standard model.

D. Many of these truncations were discovered outside the realm of string theory. For example, the 10-dimensional string was constructed using techniques that mimic more closely those used to construct the supersymmetric standard model.

II. In a previous lecture, we discussed the degrees of freedom of the graviton and the fact that these degrees of freedom can change depending on the dimension in which the mathematics is described.

A. In our world, the graviton has 2 degrees of freedom. This is given by the formula: $1/2$ of $D(D-3)$.

B. For $D = 11$, this formula yields 44 degrees of freedom for the graviton. The purpose of this lecture is to discuss supergravity in the context of 11 dimensions, but a return to our 4-dimensional world for a moment is also useful.

C. In our world, photons come in left-spinning varieties and right-spinning varieties, and their rate of spin is twice that of an electron. The number of degrees of freedom of a photon in our world is 2, found with the formula: $(D-2)$.

D. As we've said many times, the photon is a force carrier that acts between charged objects. In a hypothetical world of 11 dimensions, could there be something like the electromagnetic force that is typically carried by photons? The answer is yes, with some complications.

1. A look again at an animation of a point particle begins to show the way. As far as we have presently seen in the laboratory, things like electrons act like point particles.
2. Previously, however, we noted that, in addition to point-like objects, we can also construct "lines," called by different names, including *string* and *1-brane*. Consideration of, not a line, but a plane, leads to a *2-brane*. This can be replaced by a volume, which results in a *3-brane*. All of these objects can carry charge.
3. Setting aside strings and just thinking about particles, how do branes couple to their force carriers? To answer this, we must introduce objects called *forms*. Forms are similar to photons, except they can couple to extended objects, whereas photons can couple only to particles. The photon is an example of a form.

E. How many degrees of freedom does a form have?
1. We just discussed a sequence of objects: a point, a line, a plane, and a volume. The point was the 0-brane, and points couple to photons. The photon is also known as a 1-form.
2. A single line is a 1-brane, and these couple to 2-forms, which are similar to photons. A plane is a 2-brane, and it couples to a 3-form.
3. For every single object that can carry charge, there is a corresponding form that plays the role of the photon.
4. Now let's do some counting to find out the degrees of freedom for a form. For the 2-form, the formula for degrees of freedom is: $1/2\,(D-2)(D-3)$. For the 3-form, the formula is: $1/6\,(D-2)(D-3)(D-4)$. For $D = 11$, a 3-form has 84 degrees of freedom.

F. Remember that a graviton had 44 degrees of freedom in 11 dimensions. Given that a 3-form and a graviton are bosonic particles, it is impossible to have supersymmetry with these objects alone. Supersymmetry requires both bosons and fermions. We can see, however, that $44 + 84 = 128$. Why is this number important?

1. Remember that, in superspace, gravitons can move into one of the superdirections and become gravitinos, the superpartners to the gravitons.
   2. If there is a superpartner to the graviton in 11 dimensions, how many degrees of freedom does it have? This question was first answered in 1978 by three French physicists, Eugene Cremmer, Bernard Julia, and Joel Scherk. The answer is 128 degrees of freedom.
   3. The graviton together with the 3-form has 128 degrees of freedom. The gravitino also has 128 degrees of freedom, and supersymmetry requires a system to have a balance between degrees of freedom.
  G. This theory is called *11-dimensional supergravity*. It is not a string theory, but it uses the techniques of the standard model and Einstein's theory of general relativity in 11 dimensions.
III. Note that all the strings discussed exist in 10 dimensions, but the theory of supergravity exists in 11 dimensions. All the superstrings have truncation to supergravity and related non-string models. To what is 11-dimensional supergravity related as a truncation?
  A. This is a natural question to ask since every single 10-dimensional non-string theory can be obtained as a truncation of the vibrations of the full string. In the same way, physicists began to wonder: Is 11-dimensional supergravity the truncation of a bigger theory?
  B. Attempts to try to construct this larger theory were made in the 1980s. Objects called *membranes* were studied to see if they could be truncated and result in 11-dimensional supergravity. The problem with this program was that the modes of vibration of the membrane were infested with tachyons. It seemed, then, that 11-dimensional supergravity was not tied to a bigger theory.
  C. In 1995, Edward Witten proposed a surprising answer. He suggested that the theory of 11-dimensional supergravity was the truncation of something called *M-theory*. M-theory is truly mysterious even to theoretical physicists.
  D. The way that Witten reached his conclusion about M-theory is rather interesting.

1. He started with the type-IIA string in 10 dimensions. Then, he tried to apply the properties of the quantum world to his calculations in this theory. For example, in the quantum world, to calculate a force, Feynman graphs are added to get the quantum corrections. Witten's prescription was to try to add all possible pictures that describe a force.
2. The calculations from this process reveal that the theory in 10 dimensions resembles 11-dimensional supergravity.
3. This was the first suggestion that there was a connection between 11-dimensional supergravity and something that is extended like a string. Witten's proposal set off the third string revolution.

IV. M-theory has other implications in physics.
   A. Another illustration in which these theories are represented by dots of light serves this purpose. The dots here represent the five types of strings discussed and one dot represents 11-dimensional supergravity. Witten was able to find connections between supergravity in 11 dimensions and the closed type-IIA string in 10 dimensions.
   B. In the early 1980s, in my own research, I proposed some relationships between supergravities called *dualities*. What Witten had found was also a duality.
   C. Very rapidly after Witten, connections were found between all the strings and 11-dimensional supergravity via quantum corrections. This partially addressed the question of whether positing many different strings as a theory of everything meant that our universe had many different "everythings." If all the strings are part of a single mathematical entity, then we would have one theory of everything.

V. At this point, theoretical physicists still know very little about M-theory and are still looking for calculations to support its existence.
   A. In 1997, the most precise calculation to date in support of M-theory was proposed by Banks, Fischler, Shenker, and Susskind. We can use an allegory to help us understand the support provided by their work.

B. Imagine the most powerful microscope ever conceived. It can see well beyond what our current technology supports. This microscope might be turned to look at the string.
  1. With the increase of the magnification, we find that what we thought had seemed a single strand, in fact, looks more like a strand of pearls.
  2. This is exactly what happened with the mathematical investigations of Banks, Fischler, Shenker, and Susskind. Their calculations showed that the string is made up of something that looks like 0-branes, or particles.

C. The beads on the strand of pearls had appeared in physics earlier as *partons*. They had been discussed in the literature when physicists were struggling to understand the structure of protons and neutrons.

D. Partons are calculationally useful; that is, there are experimental physicists to this day who use the parton idea to calculate the properties of nuclear matter. However, partons don't exist on the fundamental level; they were displaced by the notion of quarks.

E. In the calculations of 1997 involving 10-dimensional Yang-Mills theories, partons were found. Furthermore, the calculations supported the idea that partons were the beads in the string of pearls that seemed to describe the string. In other words, under the string itself may lie an even more fundamental object, what is now called an *M-parton*.

F. We seem to have traveled in a big circle. We began these lectures by saying that particles do not allow us to reconcile the laws of the quantum world with the laws of general relativity and gravity. We then saw that replacing particles with filaments offers a possible resolution to this problem. We now find that our filaments may actually be like strands of pearls and that the pearls themselves are particle-like structures.

G. M-theory continues to be less than well understood. The 1997 calculations work only in a construct called the *infinite-momentum frame*. Thus, there is still some degree of controversy about whether M-theory is correct.

Readings:

Greene, *The Elegant Universe* and *The Fabric of the Cosmos: Space, Time and the Texture of Reality*.

Kaku, *Hyperspace: A Scientific Odyssey through Parallel Universes, Time Warps, and the 10th Dimension*.

Kane, *Supersymmetry: Unraveling the Ultimate Laws of Nature*.

Questions to Consider:

1. What does 11-dimensional supergravity tell us about the degrees of freedom of the graviton and the gravitino?
2. Describe the circle we've traveled thus far in our exploration of string theory.

Lecture Twenty-Three—Transcript
Can I Have That Extra Dimension in the Window?

If you listened to the last lecture, you found an explanation for why, in the middle to late 1990s, most physicists would have told you—those at least interested in the topic—that there were five different strings. There's a little bit of a problem with that because you see, strings supposedly describe everything. So how can there be five different "everythings"? There was a slight embarrassment, to say the least. There was another embarrassment, however, that was around at the time. A few people worried about it. This has to do with a theory that is not a string theory, but closely related to it. Before I come to that embarrassment, let's return, however, to the notion of a mode, because we want to make sure we have a very clear understanding of that.

If you look at the transparency I'm running, or the animation, you'll see that there is a closed string, and it's vibrating in such a way that there are three lobes that appear and disappear. You could say well, gee, why only three, why not more? The answer is that you can make the string vibrate with more. You can have a picture—instead of having these three lobes appearing and disappearing, you have five lobes appearing and disappearing, or six, or any number that you want. In fact, you could have less. Instead of having these three lobes that appear and disappear, you could have two lobes that appear and disappear. The number of lobes that you see in this thing is what we mean by a mode of vibration. This is actually not the simplest mode of vibration. This is what we might call the first excited state. The simplest one would be where there would only be two lobes that appear and disappear as the string vibrates.

Why was that relevant? When you have a string, of course, all of these are possible. You can have two lobes; you can have three, four, five, six, 17, 137, 2,000,000,618, 3.6 billion. All of those numbers could appear, and they're all possible ways in which the string can vibrate. Remember, every time the string vibrates in a different way, from our perspective it looks like a different particle. Therefore, rather than talking about the entire string, we can consider only its simplest mode of oscillation. The simplest mode would be the one where only two lobes would be appearing in this picture. It turns out that, if we restrict ourselves to this simplest way of vibrating, we're no longer talking about a string; we're only talking about part of a string. Now let's do this to the strings that were known in 1994. Remember, there were five of them. One of the interesting properties of these five strings is that they are most simply described if the world has nine spatial directions and one temporal direction. That's what we mean by a D=10 string or superstring.

In 1994, these were what were known. In the last lecture, we talked about what these names meant. Instead of taking the entire string, we now want to truncate the string and only consider its lowest mode of vibration. That's the two lobes appearing and disappearing. When you get that truncation accomplished, you're no longer talking about a string. You're talking about a particle theory because remember, the ways of vibrating correspond to particles. If you take the open string and do this truncation, you wind up with what we call ten-dimensional supersymmetric Yang-Mills Theory. If you were to take the closed type I string and perform this truncation, you get ten-dimensional N=1 supergravity plus super Yang-Mills Theory. If you were to take the closed type IIA string and perform the truncation, you get ten-dimensional N=2A supergravity. If you take the closed type IIB superstring and truncate, you get D=10, N=2B supergravity. Finally, if you take the heterotic string and perform the truncation, you again get ten-dimensional N=1 supergravity plus super Yang-Mills Theory.

There are some slight differences, however, between Case B and Case C. Now we've left behind the world of strings, and we're at the level of studying the kinds of equations that describe the Standard Model. These are the same equations that we use to describe the quarks and the neutrons, muons, gluons, photons and the entire zoo that constitutes the Standard Model. These are the same equations, except in these truncations, we're in ten dimensions, where our world is four-dimensional. One of the interesting things about these truncations that I've told you is that many of them were discovered without knowing anything about string theory. For example, the ten-dimensional string was actually not constructed using string theory, but by some other techniques that more or less mimic the way that we construct the Standard Model.

In a previous lecture, I talked about the fact that the graviton has a number of degrees of freedom that change in the dimension in which you're describing the mathematics. In our world, the graviton has two degrees of freedom. It's sort of like the photon. It can spin in the left-handed sense, or it can spin in the right-handed sense, but it spins at four times the rate of the electron. Yet it still only has these two states. The numbers of degrees of freedom in the graviton are given by the formula which says take one-half, multiply times the dimension in which you are trying to make the description, and then multiply that yet again by that same dimension minus three. You plug in D=4 here; you'll find out that that leads to two degrees of freedom, the two ways in which the graviton can spin in our world.

For D=11, you can also substitute into this formula. If you do that, you find that a graviton in a hypothetical world of 11 dimensions has 44 degrees of freedom. The reason why we bring this up is we're going to try to discuss supergravity in the context of 11 dimensions. Let's go back to four once again. In our world, we mentioned that photons can also come in left-handed spinning varieties and right-handed spinning varieties, although the rate of their spin is actually twice that of the electron. There are still only two varieties. So the degrees of freedom of a photon in our world are two. Again, if you write the mathematical equations for a photon, nobody tells you what dimension in which the photon ought to exist. Therefore, you can ask the question, "What is the number of degrees of freedom for a photon not in our world, but in an arbitrary world?" You will find by some methods of analysis that the photon for an arbitrary hypothetical world of D dimensions has D minus two degrees of freedom.

The photon, as we've talked about in many, many lectures actually is a force carrier. Typically, it carries forces between charged objects; in our world, between electrons or between quarks or anything that possesses electrical charge. That force of repulsion that charged objects acquire is due to the exchange of photons. In this hypothetical world of 11 dimensions, let's ask could there be something like the electromagnetic force? Well, the answer turns out to be yes, but there are some complications. Let's remember that as far as we can tell, things such as electrons behave like point particles. We've seen point particles. Here's our little movie of our bouncing point particle. We can think of it as an electron.

In a previous lecture, however, we have also discussed the fact that, in addition to point-like objects, we can construct essentially things that are lines. We call them by many different names. We've called them strings because a string effectively is a line. We've also called this object a one-brane. One-branes can be charged. If we consider not a line, but instead a plane, we then reach what's called a two-brane. A two-brane can be charged. We don't have to stop with two. We can let this line grow thick, and now we have a three-brane. All of these objects can carry charge. Since they all can carry charge, we can ask ourselves a natural question, namely—forget about strings—when you just talk about particles, how do these things couple to their force carriers? The answer turns out to be something that was worked out in 18th century mathematics, once again. You have to introduce objects that are called *forms*.

Forms are pretty much like photons, except that they can couple to these extended objects, whereas the photon, in fact, directly cannot. We can ask ourselves, for these forms, how many degrees of freedom do they have? For the form case, we have to be a little bit careful because remember, I took us through a sequence of objects. We started with a point. Then we went to a line. Then we went to the plane. Then we went to a volume. The point was the zero-brane. Points couple to photons. Photon is also known as a one-form, so points couple to one-forms. If we take a single line, that's a one-brane. One-branes, in fact, couple to objects that are not photons, but are very similar to photons, things that are called two-forms. Remember, we said that there's this mathematics that was developed in the 18th century, and this is what it's good for in terms of physics.

If we take a three-brane, it actually couples to a three-form. So for every single object that can carry charge, there's a corresponding form that plays the role of a photon. Now let's do some counting, because we want to look at this issue of the degrees of freedom. Remember, I kept using the word brane with a number attached. I said a point particle is a zero-brane. A line is a one-brane. A plane is a two-brane. If I make the plane thick to get a volume, it becomes a three-brane. For each of these branes, we can ask what the degrees of freedom are. I have a little table here where I have worked it out for the case of the two-form and the three-form. Remember, we're interested in this problem not necessarily in our world, but in a world where there may be more dimensions, a world where D controls the number of dimensions.

If that's the case, what we find is that a two-form has one-half the dimension in which we're interested minus two, multiplied by the dimension minus three. There are degrees of freedom of a two-form.

On the other hand, the degrees of freedom of a three-form are one-sixth times the dimension in which we're interested minus two, times the dimension in which we're interested minus three, times the dimension we're interested in minus one. What's interesting to us is this case of the three-form. Because if you plug the number $D=11$ into this mathematical equation, it tells you that a three-form has exactly 84 degrees of freedom.

Now we have to actually go back a moment because this, after all, is taking us a bit far afield, and remember that the graviton had, in 11 dimensions, 44 degrees of freedom. A three-form and a graviton are bosonic particles, so you will never have supersymmetry with those objects. To have supersymmetry, you have to have bosons and fermions. But the sum of the numbers 44 and 84 add up to 128. Why is that 128 important? You'll remember that in superspace, gravitons can move into one of the super directions and become gravitini or gravitinos, the superpartner to the graviton. If we therefore imagine looking at an 11-dimensional space, and if we start with a graviton and allow it to wander into one of the Grassmann directions, it's no longer a graviton; it's the superpartner to the graviton.

Now we can do something rather interesting. We can ask, if there is a superpartner to the graviton in 11 dimensions, how many degrees of freedom does it have? This question was first answered in 1978 by three French physicists, Cremmer, Julia and Scherk. If you ask the question, the answer turns out to be the superpartner to a graviton in 11 dimensions has exactly 128 degrees of freedom. If you've been following our discussion of supersymmetry, all sorts of bells should start to go off now because the graviton and the three-form had numbers of degrees of freedom that added up to the number 128. We have no found that the superpartner to the graviton has degrees of freedom that are 128. Therefore—remember supersymmetry requires balanced numbers of degrees of freedom—here's a system that has exactly balanced numbers of degrees of freedom. This theory was actually constructed in 1978, and it's called 11-dimensional supergravity.

Remember, this is not a string theory. What we've done is use the techniques of the Standard Model, but we use those techniques in a space of 11 dimensions. But we haven't allowed for any strings. There's actually something interesting about this because all of the strings we've talked about exist in ten dimensions. This theory of supergravity that we've talked about exists in 11 dimensions, one dimension higher. So for many years, it was a question, what is 11-dimensional supergravity related to? Remember, we went through the truncation process. Every single ten-dimensional theory, instead of looking at all the vibrations of the string, can be truncated to its lowest vibration. When you do that, you get a theory of particles. This 11-dimensional supergravity that was constructed, people began to wonder, was it the truncation of a bigger theory?

There were attempts to try to construct this bigger theory in the 1980s. There are objects called membranes, and people tried to study these things and then truncate them to show that they were 11-dimensional supergravity. There was only one problem with that program. If you take a membrane theory, and you look at all of its nodes and modes of vibration, you find out that it is infested with tachyons. From these lectures by now, we know that whenever we hear that word tachyon, it strikes fear in the heart of theoretical physicists, because you can't have probabilities that work if there are tachyons around. So this program died in the 1980s, and people said well, gee, 11-dimensional supergravity, whatever it is, is not tied to some kind of bigger theory. It was just thought to be on the side of curiosity.

In 1995, Edward Witten proposed a surprise answer. He suggested that this thing that we call 11-dimensional supergravity was the truncation of something bigger. He named this something bigger M-theory. M-theory is truly mysterious, even to theoretical physicists. In fact, maybe the M stands for mysterious. We're not completely sure why Witten gave it this name. There are lots of explanations. Some people say it's magical, others mysterious, and a couple of other explanations. It's nice to think that maybe this other thing exists and can truncate to 11-dimensional supergravity, and Witten made a conjecture that this was the case. The way that Witten reached this conjecture was rather interesting, however.

Remember, in ten dimensions, we know lots of strings. We know what those theories are, as large theories and their truncations. Witten actually proposed the following suggestion. Start with the type IIA string in ten dimensions. We know what that is. We also know that the world of string theory is intrinsically quantum mechanical. So he said calculate with this theory, but use the properties of the quantum world. In particular, remember that in the quantum world, when you want to calculate a force, you start with a picture. Then you start adding lots and lots of other pictures to get the quantum corrections. Witten's prescription was add not just many pictures, but tries to add all possible pictures that describe a force. When you do that, the calculations indicate, remarkably enough that the theory in ten dimensions with which you began starts to look like 11-dimensional supergravity.

This was the first suggestion that there's an actual connection between 11-dimensional supergravity and something that's extended like a string. This set off the third string revolution, as it's called, with many, many physicists trying to find a way to see if string theory was really related to something bigger. Let's talk about this more globally, because M-theory turns out to have other implications. We will illustrate these theories as dots of light, so there they are on the screen. There is the heterotic string up in the upper right-hand corner of my diagram, closed type I, open type I, closed type IIB, closed type IIA. Then in the upper left-hand yellowish dot, we have 11-dimensional supergravity. I already told you that Witten was able to find connections between the supergravity of 11 dimensions and the closed type IIA string in 10.

In the early 1980s, in my own research, I proposed some kinds of relationships between supergravities that were different. These things are called dualities. The thing that Witten had found was also a type of duality. When you give physicists a good idea, we're going to obviously use it to go look for other connections. Very rapidly after Witten's observation, connections between all of the strings and 11-dimensional supergravity, via quantum corrections, were found. This was actually not a bad thing because remember, we often think of strings as a theory of everything. But if there are many different strings, then does that mean that there are many different everythings? That's clearly unsatisfactory, sort of, from a philosophical viewpoint. On the other hand, if all of the strings are part of a single mathematical entity, then it makes sense that you have a theory of everything, because you've, in fact, covered all possibilities.

M-theory, in fact, was enthusiastically embraced by the theoretical physics community. In fact, we have been studying it ever since. The unfortunate thing about the study, however, is we still know very little about M-theory. Because of this situation, it's always unsatisfactory for theoretical physicists to have conjecture in place of calculations because one of the things that we constantly do, when discovering these new mathematical objects, is try to find calculations that support or can falsify their existence. Remember, physics is really not about finding the truth. What it's really about is being less wrong. If you believe something and it's falsifiable, then it's science, and we can quantify that and make progress in our understanding by removing those parts of our belief that are not correct. So, here's this proposal for M-theory. Is there a way to prove it as falsifiable, or is there a calculation that supports its existence?

In 1997, such a calculation to support its existence—in fact, the most precise definition of M-theory that we've seen—came from four physicists Banks, Fischler, Shenker and Susskind. The way that they found this support, we can understand by a little parable. I'll show this as a movie. Let's imagine that we had the most powerful microscope that was ever conceived, so powerful that it could see well beyond anything that our technological capability supports. Right now, with our most powerful microscopes, we can see atoms, but we can't see things much smaller than that. In the world of imagination, I want a super duper microscope. Let's use this microscope and look at the string. We get the eyepieces in shape, and there's the string. Sure enough, it's a string.

Since this is the most powerful microscope that we conceived, let's turn up the gain and see what happens. It still looks like a string to me. Let's continue increasing the resolution of the microscope and its resolving power, looking at smaller and smaller structures. Guess what? The thing that we thought was a single strand, in fact looks more like a strand of pearls. This story turns out to be exactly what happened with the mathematical investigation by Banks, Fischler, Susskind and collaborator. They carried out the calculations, and what the calculations showed was that if you start with a zero-brane—remember, a zero-brane is a particle—when we turned our powerful microscope onto the string, what we saw were little beads. Beads are about the same as particles.

There was a calculation that was performed by, as I said, Banks, Fischler, Susskind and Shenker, where they, with mathematics, went exactly through the process that we did with our story of the microscope. At the end of the day, what they found was a sort of bead. This bead had actually appeared in physics before; it's called a parton. Partons were discussed in the physics literature when people were struggling to understand the structure of protons and neutrons. In fact, one of the main proponents of partons back in the 1960s was Richard Feynman. So it's an idea that had been in the physics literature for 40 years. Partons, however, have an interesting life. On the one hand, they turn out to be useful for calculation purposes; that is, there are experimental physicists to this day who use the parton idea to calculate properties of protons and neutrons and nuclear matter. Guess what? It works, so partons certainly are useful.

On the other hand, what is a parton? Partons don't exist at a fundamental level. They were displaced by the notion of the quark. But still, as I said, useful for purposes of calculation.. In the calculation of 1997, looking at the ten-dimensional Yang-Mills Theory, partons were found. Furthermore, the calculations supported the idea that these partons were the beads that appear in the string of pearls in our story. In other words, that under the string itself, there may be a more fundamental object, and these would be what are now called the M-partons. Wait a minute. That means we just went in a big circle. We began these lectures by describing how particles don't provide us a way to reconcile the laws of the quantum with the laws of general relativity and gravity. There are calculation supports that suggest that if you get through the idea of particles and replace them by filaments, there is a possible resolution.

We now find that our resolution, however, when looked at by certain mathematical tools that look like microscopes, tell us that these strands are actually strands of pearls, and that the things that play the role of the pearls themselves are particle-like structures. It's a very strange circle to go in. Now do we believe it? Well, M-theory continues to be less than well understood. The calculation I described for you in 1997 makes the suggestion that this is a consistent picture. There's only one problem with it. That calculation only works if you're looking at what's called the infinite momentum frame. That's a kind of approximation to saying things are moving very fast. We don't know how to do the calculations any other way. Because of that, we cannot be completely sure what would happen if we would calculate at a velocity like something less than the speed of light or at the speed of light.

There is still controversy, or some degree of controversy, about whether M-theory exists. Let me tell you, however, another way of thinking of M-theory, another set of calculations. In 1999, two physicists, Pengpan and Pierre Ramond—now the name Ramond has appeared several times in this lecture.; this is the man who first added spin to the string in the 1970s, he's still doing physics today—but using one of the groups of Cartan, what these two physicists found is that there are an infinite number of copies where—remember, we started with 11-dimensional supergravity—we have the graviton, the three-form and the gravitino. They have found that there are an infinite number of copies where the numbers of degrees of freedom exactly balance, and that would be the condition you need for M-theory. Therefore, M-theory is probably out there. It's waiting for some incredibly beautiful idea before we can see its full beauty.

Lecture Twenty-Four
Is String Theory the Theory of Our Universe?

Scope:

I hope that these lectures have shown you that string theory is embedded in the fields of science and mathematics as we have known them for hundreds of years. String theory weaves together an amazing story with strands contributed by both mathematicians and physicists, some of whom are long dead and some of whom are working today. We know that string theory encompasses a large collection of mathematical facts, but we have not yet seen evidence of it in the laboratory. We are still looking in string theory for the physical insight that will be the equivalent of the "Man in the Elevator" that led Einstein to the theory of general relativity. My own research today is aimed at finding the denizens of supersymmetry in the hopes of contributing a new tool in physics for the solution to this problem.

Outline

I. String theory is a collection of non-trivial, mutually enforcing mathematical facts, but we have not yet seen evidence of it in the laboratory.

  A. For me, string theory is one of the most remarkable concepts I've seen appear in the realm of applied mathematics. The structure of superstring/M-theory touches on a number of subjects in mathematics: algebraic geometry, calculus, differential equations, differential geometry, Lie group theory, number theory, topology, and many more.

  B. Amazingly, this construction, which started in physics, has started to "eat" mathematics. Before string theory, the last intense conversation between mathematics and physics took place in the 1930s, driven by the discovery of quantum physics.

  C. Often, physicists "invent" mathematics they need to describe experiments. From a mathematician's viewpoint, this method lacks rigor. Mathematicians often find themselves in the position of taking the suggestions of physicists and turning them into real mathematics.

II. One way to define the word *serendipity* is "the occurrence and development of events by chance in a happy or beneficial way."
   A. The word was coined by Horace Walpole and drawn from a story entitled "The Three Princes of Serendip."
   B. Superstring/M-theory has an extremely high S.Q.—serendipity quotient. The mathematical accidents that must occur in order for string theory to be logically rigorous are incredibly large in number. In fact, many researchers call the theory "string magic."
   C. In physics, when we see an enormous number of mathematical accidents, we believe we are looking at mathematics that is tied to the real world. An example of this is when Dirac tried to write equations to describe the electron in a way that was consistent with Einstein's rules of special relativity. He succeeded, but he also found that his equations demanded the existence of an antiparticle for the electron.

III. Why did we push physics all the way to strings?
   A. The idea of atoms seemed good enough at one point, but when we start probing the atom, we find that it contains elementary particles.
   B. We've studied elementary particles for 60 or 70 years. Literally tens of thousands of pieces of data agree with the mathematical description of elementary particles, called the *standard model*. But none of the experiments we've performed with elementary particles takes into account gravity. If we try to write a theory combining quarks, leptons, force carriers, etc., with gravity, the result is mathematical nonsense.
   C. After creating the theory of general relativity, Einstein spent 30 years trying to find its successor, which he called the *unified field theory*. Einstein was disturbed by the fact that his calculations could be made in such a way as to describe a universe in which only space and time existed, with no matter.
   D. String theory is fundamentally different from general relativity in that the equations of string theory absolutely require matter.

IV. String theory is not a complete story, nor is M-theory.
   A. In the quantum world, we cannot distinguish between particles and waves, and the way to describe a wave is with the mathematical tools of a *field*. In these lectures, I never mentioned a string field because the field is the boundary of our knowledge in superstring/M-theory.
   B. A field-theory description for the open string was discovered by Witten in 1984–1985. For closed strings and the heterotic string, no one knows how to write field-theory descriptions, which will be the true analogs of Einstein's theory of general relativity. Why can't we find them?
      1. In 1905, Einstein had his miracle year, but in 1907, he noticed workers on a roof opposite his office building and had a thought: If one of the workers should fall, he would not feel his weight.
      2. Einstein later described this as the "happiest thought of his life," because it led to general relativity. General relativity is the genesis of the Big Bang.
      3. Notice that this theory didn't come through mathematics but through a deep understanding of one aspect of the universe.
      4. In an animation of Einstein accelerated in an elevator, the acceleration feels like the force of gravity to him. If a ball were thrown into the elevator from the side, it would appear to be falling. Using this same reasoning with a beam of light, Einstein concluded that light must be bent by gravity.
      5. This story of "The Man in the Elevator" can be used mathematically to derive Einstein's equations of motion. But we have no such story for string theory. Even though we have a vast collection of mathematical facts, we do not have a point of physical insight that tells us why string theory must be correct.
      6. In this way, we can think of string theory as "The Search for the Missing Man in the Elevator."
V. Superstring/M-theory almost makes some predictions.
   A. String theory tells us we should expect ordinary matter to have superpartners. But finding the superpartners will be only indirect evidence of string theory.

- **B.** Looking at the running of the coupling constants in experiments we can perform in the next decade, we may be able to detect supersymmetry.
- **C.** We may also find a boost for superstring theory by studying the fluctuations in the temperature of the cosmic microwave background.
- **D.** Whether or not we find superstring theory as a physical principle, it seems secure as a mathematical theory.

VI. As an active researcher, I am aware of the problems presented by string theory.
- **A.** We need to find for string theory the equation that plays the role of Einstein's equation in general relativity. By solving other problems in string theory, we might make a contribution to solving this big problem. My current research is aimed in this direction.
- **B.** Let's look at a picture of SUSY. The two blue areas in this picture represent universes that contain only bosons, on the one hand, and only fermions, on the other. For the past five years or so, I've been studying the mathematics of this picture.
- **C.** The arrows that point from the blue area on the left to the blue area on the right and vice versa represent the supersymmetry transformations. What's in the blue regions? This brings us to a part of mathematics known as *representation theory*.
- **D.** When we talked about spinors and vectors using the analogy of arrows and balls, those were representations. For rotational symmetries, we know every single representation that occurs. But we do not know all the representations for supersymmetry.
- **E.** By studying the mathematical questions associated with these representations, I hope to answer a fundamental question: What's in the zoo of supersymmetry? What's the complete list of denizens that inhabit the blue regions in our picture?
- **F.** What does this research have to do with string theory? We know that strings vibrate in all possible ways; thus, we

expect that we will find, in string theory, all possible representations.

G. My colleagues and I haven't solved this problem, but we have found interesting connections to other fields of mathematics, such as K-theory. We have found pictures of the representations in these blue regions, suggesting that we are reaching a deeper level of understanding.

H. If we're successful in our efforts to find all the denizens, we will turn our focus back to the string and ask whether the equation that plays the role of general relativity for string theory can describe all these possible ways of vibrating.

VII. Here are some of the denizens we've found in the world of supersymmetry.

A. Michael Fox and I, by studying the mathematics of these blue regions, remarkably found that there are simple pictures that contain the information that describes these representations.

B. We chose the name *Adinkra* for these new denizens, from the African language Ashanti, defined as a symbol carrying hidden meaning.

C. These Adinkras have a number of interesting properties; for example, they can be folded up, similar to proteins. At this point, we are simply classifying the number of representations we have found.

D. In closing, we see a picture of one of the most interesting beasts my colleagues and I have met so far. The complicated figure on the left is a picture of the force carriers and their superpartners. In other words, in the blue regions, we have been able to identify the representations that have exactly the properties of the gluons and gluinos, photons and photinos, and so on. The figure on the left is a picture of the quarks and squarks, electrons and selectrons, and so on.

E. We hope that this graphical approach will allow us to find all the inhabitants of the blue regions. Should we do so, physics will have a new tool with which to study the mysterious world of M-theory.

Readings:

Walpole, *Horace Walpole and His World: Selected Passages from His Letters*.

Questions to Consider:
1. What is the most significant problem facing string theory?
2. What are some of the discoveries that may give a boost to string theory?

Lecture Twenty-Four—Transcript
Is String Theory the Theory of Our Universe?

We've come to the final lecture. For those of you who have stuck through the grind, I hope that what we've done is really to show you how string theory, this mysterious thing that you may have heard so much about, also called Superstring/M-theory, is embedded within the fields of science and mathematics as we've known them for hundreds, if not thousands, of years. String theory weaves together an amazing story where strands are contributed by researchers in both mathematical and physical sciences who have been long dead. Yet the results that they left behind suddenly illuminate something that's as complicated as string theory. String theory, as I have tried to emphasize in these lectures, is a collection of highly non-trivial mutually enforcing mathematical facts. It's not yet science. We've not seen the evidence of it in the laboratory.

I judge personally that this is one of the most remarkable concepts that I've ever seen appear in the realm of applied mathematics. The structure of Superstring/M-theory touches on incredibly large numbers of subjects in mathematics. Some of these include algebraic geometry, calculus, differential equations, differential geometry, Lie group theory, number theory, topology, monodromy and many, many more. It's amazing that this construction, which starts in physics, actually has started to eat up mathematics. What do I mean by that? Normally, as you watch the process of physics and mathematics, they're not always talking to each other, though periodically, they do.

Before string theory, the last time where you could find in the literature of both disciplines an intense conversation was in the 1930s. This was driven by the discovery of quantum physics. The physicists knew what they needed in terms of mathematics to describe the world of the small, and physicists, believe it not, were not always so fastidious in the use of mathematics. In fact, if you watch physicists use mathematics, they just make it up. They need a particular mathematical widget, tool or device, and they say well, this is what I want. I'll set a few rules down. I'll calculate with it. If it produces a result that agrees with an experiment, physicists are perfectly happy.

This is not the case for our mathematical colleagues. Oftentimes, it is the case that when physicists produce mathematics, from the point of view of real mathematicians, there's a lack of rigor. Therefore, mathematicians are often at the point of taking the suggestions that physicists have and turning them into real mathematics. I've had some experience with this process recently because I will have a chance, I hope, in a little while to tell you about what I'm doing right now. I'm an active researcher. Although I enjoy teaching and I enjoy lecturing, there is an enormous sense of enjoyment in entering the world of mathematics and going on a hunt for an idea.

One way to define the word *serendipity* is the occurrence and development of events by chance in a happy or beneficial way. The word serendipity was created by H. Walpole, and it's based on a story called *The Three Princes of Serendip*. The book is about a land in Asia where there are three princes who are constantly getting into trouble, but they get out by the most remarkable means. They're not wise, they're not clever, but they have lucky accidents. So the word serendipity itself is serendipitous because it was coined from the story. When you look at something such as Superstring/M-theory, it has an incredibly high SQ, that is, serendipity quotient. The mathematical accidents that must occur in order for string theory to be logically rigorous are incredibly large in number. In fact, so much so that many of the researchers in string theory call it "string magic" because things occur that you just couldn't expect.

It's been a pattern in physics that, when enormous numbers of unlikely mathematical accidents occur, you're looking at a piece of mathematics that is tied to the real world. We can see other examples of this. In a previous lecture, I've talked about Dirac. He was trying to write an equation which would describe the electron, which would be consistent with Einstein's rules of special relativity. He succeeded. However, he found something in the equations he had never put in, namely, the equations demanded the existence of an anti-particle to the electron. It's a mathematical accident. Nonetheless, we now know that that accident is a part of reality. So this is what we mean by serendipity in the mathematics that describes physics. It's incredibly high in string theory, and it is, in fact, a characteristic of good theoretical physics.

Why did we push all the way to strings? Because you might, at any stage, have asked the question, well, why don't they stop? After all, atoms seemed good enough at one point. That's true until you start probing what's in the atom. When you take an atom and shake it, and you hear something rattling inside, you have to worry about what's that something. That takes us to the world of elementary particles. Elementary particles, we've been studying for roughly 60 or 70 years. Literally tens of thousands of pieces of data agree with the mathematical description of elementary particles. That description is called the Standard Model. You might say well, why don't they stop? Once again, there's a problem with the Standard Model. You see, all of these experiments that we describe and are able to perform do not take gravity into account. If you try to write a theory which includes the description of quarks and leptons and the force-carrying particles and include gravity, you have a mathematical piece of nonsense; that is, you have something that makes no prediction at all. That's why we couldn't stop, because gravity, most apparently, is around us. So why should not the elementary particles also be subject to gravity?

Einstein, after creating the Theory of General Relativity, spent 30 years trying to find a successor. Part of the reason for that was his acute sense of what is right. If you read about Einstein—and I've been fortunate enough this year to be forced to read about Einstein, because it's been a topic of much discussion in this Einstein World Year Physics—you find out that Albert Einstein was an exceptional physicist working in an exceptional manner. Not at all what you might think, that is, he didn't start with the mathematics. Typically, Einstein would start with a story. It's this fine sense of what is right that drove Einstein to seek the theory that he called the Unified Field Theory. How did he reach this sense? In his equations describing the universe, the equation, as do all equations, had two sides. His equation has one side that describes the curvature of space and time. The other side of the equation is where you put matter.

The thing that disturbed Einstein so much was the following. You could take matter out of his equation, so put a zero on that side of the equation. The other side of the equation works just fine. Therefore, what you wind up describing is a universe where there is only space and time, but nothing else. In a theory that is truly fundamental, something is wrong there. Because after all, in our universe, it's not just space and time we're around. It was his sense that, at a fundamental level, whatever equation was correct, it ought to describe more than just space and time. His equations describe a grand stage, but that stage may be empty. A more complete theory would describe the stage, talk about the actors that have to be on the stage, and even perhaps talk about the play they must perform. That's the sense that's driving him to seek the Unified Field Theory.

String theory is fundamentally different from Einstein's view as expounded in General Relativity. In string theory, if you take the four-dimensional versions, you have no mathematical choice as to whether matter exists. You must have matter, or the equations don't make sense. That is somehow very deeply satisfying when you start to think about the aesthetics of physics. After all, why are we here? Well, if you have an explanation that says otherwise the universe makes no sense at all, that's a pretty good reason for our presence. String theory or Superstring/M-theory, or the variety of names by which these constructs are known, are the first time in physics when this statement could be made. String theory is not a complete story. When I got to the end of my presentations, I was talking about M-theory, but M-theory is not the end. Why not?

If you look very carefully in my presentations, you should have caught onto a strange dichotomy. I started talking in the language of elementary particles about particles, but then I talked about the force carriers. At some time, you may have heard the world field theory slip in. That's because, in fact, in the world of the quantum, we cannot distinguish between particles and waves. The way that you describe a wave is with the mathematical tool of the field. In all of my presentations about string theory, you probably never heard me talk about a string field. The reason that I didn't talk about it is because that's the boundary of our knowledge in Superstring/M-theory. For the open string, there is a field theory description. That means something that is more general than thinking about filaments. It was written down by Edward Witten around 1984 or 1985. For the closed strings and the heterotic string that I have described for you, no one in the world knows how to write its field theory. Why is that important?

The field theory versions of these discussions are the analogs, the true analogs, of Einstein's Theory of General Relativity. So why can't we find them? Well, let me talk about how Einstein found his equations. In 1905, Einstein had the year of miracles. However, in 1907, he was still a patent clerk. One day he looked out of the window of the office and noticed across the street some workers working on a roof. He had a thought. If one of them should fall, he would not feel his weight. This is, in fact, what shows the exceptional way in which Einstein worked because, you see, he described this scene himself later around the 1920's, 1923 or so, and he called this the happiest thought of his life. The reason he referred to it this way is because it was this single thought that led to the theory we call general relativity.

General relativity is the genesis of the Big Bang. It's a theory that begins to describe the entire history our universe, but it starts with a story. It starts with a parable, and that's how Einstein worked. It doesn't come first with the mathematics; it comes with a deep understanding of an aspect of the universe. We can see this deep understanding in a cartoon. If you look at this, you'll see we have a cartoon Einstein. He starts to be accelerated, and as he's accelerated, the acceleration feels like the force of gravity to him. If a ball were to travel from the side, since the elevator is accelerating upward, the ball would appear to fall. By using this same reasoning with a beam of light, Einstein concluded that light must be bent by gravity because it would fall in this situation. He said you couldn't tell the two apart. This is the story that's often called "the man in the elevator."

The interesting thing about the story of the man in the elevator is that you can take that story, if you have a sufficient amount of mathematical training, and use that story alone to derive Einstein's equations of motion, his Theory of General Relativity. We have no such story for string theory. That means, even though we have this incredibly large collection of mathematical facts, we do not have the point of physical insight that tells us why string theory must be correct. That's the thing that's lacking. That's the boundary that perhaps one day—certainly many of us hope—someone will have a bright idea like Einstein's story about the man in the elevator. In fact, I often refer to string theory or Superstring/M-theory presently as the search for a missing man in the elevator because if we have that story right, we will have made an enormous leap in terms of physically understanding why these theories must be true, not because of mathematical consistency.

As I said, string theory is not at its end. It's at sort of a midpoint. We're still trying to push hard to create it. Superstring/M-theory almost makes some predictions. It tells us, for example, that we should expect ordinary matter to have superpartners. These superpartners perhaps are on the order of 300 to 1,000 times as heavy as the average of the ordinary matter that we know. If we find the superpartners, they are only indirect evidence of string theory. We can also say that string theory, because of the presence of supersymmetry, tells us that by looking at the running of the coupling constants in experiments that we can perform within the next decade, we ought to at least be able, perhaps, to detect supersymmetry.

Perhaps it will come in the form of looking at something such as the cosmic microwave background where, by studying the fluctuations in the temperature of the background, we can see, perhaps, the signature of superstring theory written in the heavens. Whether we find superstring theory or not as a scientific principle, I'm pretty convinced that as a mathematical theory, it's probably here to stay. Maybe it will make it as a description of the universe, maybe not. But in terms of mathematics, there's nothing that I know in the history of mathematical physics that comes even close. In the worst possible case, we can probably say that Superstring/M-theory will become a branch of mathematics.

What am I doing? Well, I'm aware of these problems; most of my community is. Our holdup, our difficulty, is we don't know how to move past these problems. Genius does not show up on schedule. Bright ideas cannot be commanded into existence. Most of the time, as theoretical physicists, we're confused. We're trying to find correct answers to solve problems. Let me give you some insight into the things that are of interest to me right now. I talked to you about the big problem of Superstring/M-theory. That big problem is to find for string theory the equation that plays the role of Einstein's equation in general relativity. That's the big problem. No one in the world knows how to solve that. That problem has been around for about 15 to 20 years.

On the other hand, there are other problems that perhaps we can make progress in and, by studying these problems, might contribute ultimately to the solution of this big problem. That's one of the things I've been doing recently. Let me show you a bit of research, current research that's underway. I'll begin by showing you a picture. That's a picture of SUSY. Of course, you might say well, what in the world does that mean? Let me talk about this picture a little bit. The two bluish areas in this picture represent universes which only contain bosons on one hand and fermions on the other. One is to imagine that the number of bosons in one is equal to the number of bosons in the other. There are mathematically precise ways to study these issues. In fact, what I've been doing for the last five years or so is to study this mathematics.

Interestingly enough, by this picture, one is led back to study the mathematics of Clifford. We met Clifford in our discussions some time ago. We'll return to Clifford. You'll notice that there are arrows that are pointing back and forth in this diagram. The arrows that go from the blue area on the left to the blue area on the right and vice versa are the supersymmetry transformations. Remember, we kept talking about taking ladders and lifting them from horizontal positions to vertical and back. I have been studying the mathematics that maps between these two regions, back and forth. In fact, we've begun to see some interesting structures emerge. One of the questions that you would want to know is what's in those blue areas? This is a part of mathematics that's known as representation theory. This is the problem I have chosen to attack. It's related to string theory, but it's not string theory itself.

We have talked many times in this presentation about symmetries and about the objects that cause rotations and the number of such objects. One thing that we haven't talked about was to ask a very simple question, what else are there besides ladders? What do we mean by that? What else are there besides arrows? Let's remember. I've talked about spinors as being an object that's a little bit like an arrow. If it's pointing at you and you walk around it once and look at it, you actually find it's pointing away. If you walk around twice, it's pointing at you, and nobody has touched it in between. That's a spinor. What's a vector or an arrow? An ordinary arrow, if it starts pointing at me, if I walk around once and look at it, it's still pointing at me.

Suppose I started with a ball. If I walk around the ball, not even all the way, if it has no features, it looks the same to me. Ball, spinor, arrow—those are what we call representations. Each one of them has a property that you can express on traveling around it. Are there more such things? The answer is yes. We know, in fact, all of the objects in this sequence that I've described. They're called representations. For rotational symmetries, which is what it means to go around something, we know every single object that occurs. Cartan, in his classification, told us about not just rotations, but all possible symmetries. Believe it or not, we understand all of their representations. When Cartan was doing his work, supersymmetry was not known. Remember, supersymmetry only appeared in the 1970s in the West.

To this day, we don't know all of its representations. By going back and studying the mathematical questions that are associated with this, it's my hope to first answer this fundamental question, namely, what is there in the zoo of supersymmetry? What's the complete list? If we think of them as denizens as we do for the quantum world, what's the complete list of denizens that inhabit these blue regions? This is a question, it turns out, that physicists have not been able to answer for 30 years. In my quest to answer this question, I've joined with other physicists. There are three of us who are mathematicians and, counting me, three physicists. We have an interdisciplinary collaboration where we are studying this question about what resides in these blue regions.

You might wonder what this has to do with string theory. String theory has supersymmetry all over it. If that's the case, we know that string theory vibrates in all possible ways. Therefore, since it vibrates in all possible ways, the expectation is, within string theory, you will find all possible representations. But remember, that's what I'm looking for because I want to know what all the possible representations are. We haven't solved this problem, but we have found connections to interesting pieces of mathematics. For example, there's a subject in mathematics called K-theory. It would take me another entire lecture course to explain to you what K-theory is. But the point is it's an interesting mathematical tool that we have begun to use to find these denizens that inhabit the blue regions.

I can show you some of them, and I will in a few moments. We've also found that there are other pictures of these objects that exist in these blue regions. The representations have pictures. At this stage, one of the things that's emerging is this whole issue of picture making. You will notice that, in this course, I have mostly been talking about mathematics, but with a minimum of mathematics. I've been diligently trying to use the pictures of the mathematics to convey the ideas that sit behind string theory. This idea that pictures can replace mathematics speaks very powerfully to an alternative way to understand mathematics. In fact, we have used this, for example, in Feynman diagrams where there's a story associated with the picture. By the use of Feynman diagrams, I hope I've opened up a point of view for you on how the quantum world works.

The fact that my collaborators and I have begun to find new pictures of this mathematics suggests that we are reaching a deeper level of understanding that has previously occurred. If we're successful in our efforts to find all the denizens, then, of course, we want to turn the focus back to the string. Because if you know all the possible ways the string can vibrate, then you can use that as an alternative point to ask, in the equation that plays the role of general relativity, can you describe all of these possible ways of vibrating? So there are things that are actually linked. Symmetries always work like that. Remember, we've used symmetry repeatedly in the course where, by knowing one piece of information of symmetry, you can derive other things.

Now let me introduce you to some of the denizens, in the world of supersymmetry this time, not in the quantum world. There are these pictures that you see before you. Last year, working with a physicist by the name of Michael Fox, by studying the mathematics of these blue regions, we were able to tease out of the regions some of the representations. Most remarkably, we found that there are simple pictures which contain the information which describes these representations. I'm showing you some of them here. This, of course, is a little bit like going to a new continent and seeing new animals. Imagine the first settlers that went to Australia to see kangaroos. Wow, a new kind of animal that has never been seen before. When you see a new animal, one of the things you want to do is to name it.

These pictures needed a name, and Michael Fox and I decided to use the word *Adinkra*. An Adinkra is a symbol that carries hidden meaning. It comes from the West African language known as Ashanti. We have found these objects which are the beginning of a list of representations that inhabit the blue regions. We have found a number of very interesting properties. For example, these things can be folded up kind of like proteins, which came as a very great surprise to us. They come in marvelous varieties. Here are some more of them. They have properties where you can turn them a certain way, and you have another sort of animal. You can switch the legs in certain ways and you have another sort of animal. I'm using the word animal here in place of representation. We are, at this point, simply classifying the number of representations, much as an explorer in a new land would begin to classify the flora and the fauna.

Let me show you one of the most interesting set of beasts we have met so far. There are these. There's a particular reason why I put these up for you, which I will come to. There are, in these blue regions, representations for which these pictures are a complete description. The one that's on the left—the reason why I put it up is you'll remember, in the supersymmetrical universe that we've been describing, coming from string theory, there are force carriers and their superpartners. Also in this supersymmetric version of the Standard Model, there are matter fields and their superpartners. It turns out that the picture that you see on the left is a picture of the force carriers and their superpartners. We have been able to identify in the blue regions the representations that have exactly the properties of gluons, gluinos, photons, photinos, W particles and winos, and what have you.

On the other hand, the picture that you see on the right here is exactly a picture of the quarks and the squarks, the electron and the selectron, and all of the rest of the spin-zero and spin-one-half particles that we have found in the Standard Model. These pictures are very new. In fact, most physicists don't yet know that these things exist. It is our hope that, ultimately, this kind of a graphical approach will be possible for us to find all of the inhabitants of the blue regions that I showed you. Should that be the case, then there will be a new tool with which to begin to study the mysterious world of M-theory. Is it going to work out? I don't know. That's why we call it research.

Timeline

1686	Newton completes the *Principia Mathematica*, which includes his laws of motion and theory of gravity.
1783	Michell reasons out the concept of a black hole.
1811	Fourier develops a general method for describing vibrating systems.
1834	Russell observes a solitonic wave while riding his horse near a canal in Edinburgh.
1844	Grassmann develops the mathematics of *anti-commuting numbers*.
1846	von Haidinger discovers that he can detect the polarization of light with his eyes; this effect is called *Haidinger's brush*.
1854	Riemann shows that geometry done on a flat plane versus any curved surface is mathematically consistent and defines the *curvature tensor*.
1869	Mendeleev designs the periodic table of the elements.
1873	Maxwell publishes his treatise containing the four equations of electromagnetism and showing that they imply electromagnetic waves that propagate at the speed of light.
1874	Stoney posits that the atom can be broken into parts.
1888	Hertz shows experimentally that radio waves exist in physical reality.

1897	Thomson conducts experiments leading to the discovery of the electron.
1900	Planck postulates that energy of light is carried in discrete packets, called *quanta*.
1905	Einstein's "miracle year," during which he publishes papers on special relativity, the photon concept with application to the photoelectric effect, and Brownian motion.
1913	Beginning of the quantum revolution with the publication of Bohr's quantum theory of the atom; Cartan writes the mathematical description of spin.
1916	Einstein publishes the general theory of relativity.
1918	Noether develops her theorem relating symmetry to conservation laws.
1919	Observations by Eddington confirm general relativity's predictions of the bending of light by the Sun's gravity.
1921	Stern-Gerlach experiment reveals that all electrons are spinors.
1923	de Broglie argues for wave-particle duality.
1925	Pauli formulates the exclusion principle and a new quantum property of the electron.
1927	Heisenberg formulates the uncertainty principle; Wigner

	introduces the concept of parity conservation.
1928	Dirac writes equations for the electron and discovers its antiparticle, the positron.
1930	Pauli posits the existence of the neutrino.
1932	Chadwick discovers the neutron.
1934	Fermi develops the theory of beta decay.
1938	Kemmer proposes the idea of isotopic charge space, which requires only angles to completely specify location.
1948	Gamow develops the Big Bang theory; Feynman, Schwinger, and Tomonaga produce the theory of quantum electrodynamics, successfully uniting special relativity with quantum mechanics.
1953	Stückelberg and Petermann show that the coupling constant in the quantum world is a function of the energy at which it is observed.
1954	Yang and Mills develop Yang-Mills theories, providing the mathematics to describe gluons, w-particles, and z-particles.
1957	Bardeen, Cooper, and Shrieffer explain the workings of superconductors.
1962	Tachyon named by Feinberg.
1964	Quark model developed both by Gell-Mann and by Zweig; Brout, Englert, Guralnik, Hagen, Higgs,

	and Kibble develop mathematics of spontaneous symmetry breaking.
1965	Cosmic microwave background observed by Penzias and Wilson.
1966	Stanford Linear Accelerator Center (SLAC) starts operation; Berezin introduces the classical description of spin by anti-commuting Grassmann variables.
1967	Electroweak unification proposed independently by Weinberg and Salam, based in part on contributions by Glashow; Penrose introduces the concept of twistors.
1968	Development of the original version of string theory based on the bosonic string; Veneziano develops the dual resonance model, later recognized as the theoretical base of the string theory version of quantum gravity.
1969	Deep Inelastic Scattering Experiment at SLAC reveals evidence of quarks inside protons and neutrons; Adler, Bell, and Jackiw identify the presence of anomalies in the quantum world.
1971	Neveu, Ramond, and Schwarz present a variation of the original string theory composition that permits the inclusion of "notes" that have spin rates that are odd integers times the rate of spin of the electron, thus moving physics from the first generation of the string, the bosonic string, to the level of the spinning string.

1972	Modern view of quarks developed by Gell-Mann, Fritsch, and Bardeen.
1973	Bekenstein suggests that black holes should have a well-defined entropy.
1974	Georgi, Quinn, and Weinberg show that the rates at which coupling constants change can be calculated if the number and types of denizens are known; Salam and Strathdee propose the concept of superspace.
1974	Georgi, Weinberg, and Quinn derive equations that describe the trajectory of *running coupling constants*.
1974–1975	Hawking verifies Bekenstein's suggestion and proposes that black holes radiate heat through quantum processes.
1975	Two groups, Ferrara, Freedman, and van Nieuwenhuizen in one and Deser and Zumino in the other, posit supergravity.
1977	Completion of the standard model with the discovery of the bottom/anti-bottom meson by Lederman; Giozzi, Scherk, and Olive eliminate the tachyon with their hypothesis of supersymmetry; Wess and Zumino and, later, Deser write a curved supergeometry, a construct that incorporates Einstein's notion of curvature into superspace.
1978	Cremmer, Julia, and Scherk propose a theory of supergravity in 11 dimensions.
1982–1984	Green and Schwarz show that the SO(32) superstring is

mathematically consistent in 10 dimensions, marking the beginning of the first string revolution.

1983 .... Experimental verification of electro-weak unification.

1984 .... Gross, Martinec, Rohm, and Harvey write equations describing the heterotic string; Kazakov shows, in the supersymmetrical model, the strong, weak, and electro-magnetic forces all unify at the same energy.

1984 .... Friedan shows string theory is the first system where quantum theory is consistent with equations of general relativity.

1984–1985 .... Witten writes a field-theory description for the open string.

1986 .... Kawai, Lewellen, and Tye show that the complicated quantum force law for gravitons can be written in the form of simpler quantum force laws for gluons.

1988 .... Gates and Siegel find the final formulation of the heterotic string, in which Kemmer variables appear. This offers the possibility of describing *four*-dimensional strings.

1995 .... Witten proposes M-theory, a single construction that encompasses all of the 10-dimensional superstrings, heterotic strings, and 11-dimensional supergravity.

1996 .... Super-Kamiokande neutrino detector begins operations in Japan.

1997 .... Maldecena discovers the AdS/CFT correspondence, the unexpected relation between theories involving gluons and a theory of gravity in a space of one greater dimension, marking the beginning of the second string revolution and increasing interest in string theory as a possible "theory of everything"; Banks, Fischler, Shenker, and Susskind perform calculations supporting the idea that the string is made up of 0-branes called *M-partons*.

1998 ....Experiments at Super-Kamiokande reveal that neutrinos have mass. Cosmic expansion of the universe is found to be accelerating.

1999 ....Randall and Sundrum propose the brane-world model of the universe.

2004 ....Gates and Faux propose Adinkra as a method of organizing the representations of supersymmetry.

Glossary

AdS/CFT correspondence: A set of mathematical prescriptions that suggests the possibility of calculating the properties of gluons and quarks by instead considering a corresponding theory involving only gravity in a mathematical description of a space with one more spatial dimension whose geometry is given by a negative constant value for its Riemann curvature tensor.

Anomalies: A charge that may be conserved while ignoring quantum effects but may not be so conserved when they are included. The non-conservation is caused by anomalies.

Anticommutativity: The property that implies that when two objects (these may be numbers, functions, etc.) are multiplied in one order, the result is the negative of the number that results from using the opposite order.

Antimatter: Composites made of the antiparticles of those found in ordinary matter. If equal amounts of antimatter and matter are brought together, each is completely destroyed, to be replaced by an amount of energy that satisfies Einstein's famous equation, $E = mc^2$.

Baryonic matter: Matter composed mostly of baryons. Protons and neutrons are the most familiar forms of baryonic matter. The distinction between baryonic and non-baryonic is important because the processes that synthesized baryonic matter are tightly constrained by the occurrence of the Big Bang.

Baryons: Particles of matter that contain triplets of quarks. The term *baryon* comes from the Greek word *barys*, meaning "heavy."

Beta decay: The process by which a neutron is transmuted via the weak interaction into a proton, electron, and anti-neutrino associated with the electron. This process lies behind the natural radioactivity of such substances as uranium.

Big Bang: A mathematical solution to the theory of general relativity that implies the universe emerged from an enormously dense and hot state about 13.7 billion years ago.

Big Ear: A microwave radio telescope that first detected the cosmic microwave background (CMB), then called *three-degree radiation*.

Binary stars: Two stars orbiting about a common center. They often appear as a single image in a telescope. Sometimes called a *double star*.

Black hole: A solution to Einstein's theory of general relativity with the property that nothing, via the laws of classical physics, can escape falling into the black hole's singularity after coming within the radius of the black hole's event horizon.

Brane: Used in the context of present-day mathematical physics to describe the collection of elementary objects that includes points, line segments, planes, etc. A point is a *0-brane*. A line is a *1-brane*. A plane is a *2-brane*. Within a given space with dimension D, there is a maximum brane, a *D-brane*, that is also called a *space-filling brane*.

Calabi-Yau manifold: A type of compactification technique applied to models with extra dimensions. The extra dimensions are assembled into three complex numbers (using the usual factor of i) and restrictions are put on them so that the surfaces they describe possess special properties.

Chan-Paton factor: Originally, the mathematical constructions that described the charges that could be appended to the ends of open strings. In modern interpretations, these mathematical objects describe the various ways that open strings can end on branes.

Charge conjugation: The act of replacing an elementary particle by its antiparticle.

Charge conservation: The fact that electric charge can neither be created nor destroyed. If the charge of an object changes, it must be because an electric current flowed to implement this change. This rule may be applied to other sorts of charges in place of electric charge.

Clifford algebras: The basic mathematical formulae, created in 1876 by William Clifford, that allow for a description of spinors.

Compactification: The concept or process by which extra dimensions, considered actually to exist, may be subjected to additional mathematical restrictions so as to render the resulting models consistent with the physics of a space without extra dimensions. The simplest example is the projection of an image in our world onto a screen. Many mathematical properties of the object in our world must have a "'shadow" description on the screen.

Computer-enabled conceptualization: The concept that computers offer a genuinely new and alternative way for humans to access mathematics to conceptualize properties and structures of the universe. Previously, this has mostly been enabled with the use of mathematical symbolic systems.

Concordance model: The current view of the universe that posits that it possesses: (1) at the instant of the Big Bang, equal amounts of energy distributed between gravity and matter; (2) dark energy that explains its accelerating expansion; and (3) dark matter that controls galaxy lifetimes. Further, the model posits that the universe is composed of approximately five percent of the type of matter that is currently known to science.

Conformal system: Any mathematical or physical system with the property that the physics it describes is the same when examined at all possible scales of sizes. This property sometimes is called *conformal symmetry*.

Cosmic microwave background (CMB): A form of electromagnetic radiation (in the form of microwaves first found in 1964) that emanates throughout all directions in the universe. It can be regarded as the universe still reverberating from its Big Bang creation. As well, the CMB is a cosmic fossil giving concrete evidence for the occurrence of the Big Bang in the same way that ordinary fossils give concrete evidence for the existence of extinct life forms from Earth's history.

Cosmological constant: A parameter or fudge factor originally introduced by Einstein into the equations that describe the properties of space and time. His original purpose for doing this was to describe a universe that is static and eternal. Today (using a sign opposite to the one chosen by Einstein), the cosmological constant is considered to be a possible source of dark energy.

Coupling constant: A parameter in nature that measures how strongly a force affects objects that carry a corresponding charge. For example, the measured electrical charge of the electron determines how strongly the electromagnetic interaction affects the electron.

Coupling constant (running of): Due to the effects of the quantum world, the value of a coupling constant changes depending on the energy at which the interaction occurs.

Covariant lattice: An approach to constructing genuine four-dimensional strings via the use of bosons that describe crystal-like lattices. The "notes" produced by these, when combined with those from a four-dimensional bosonic string, are supersymmetric.

CPT theorem: The mathematical and experimental fact that combining the effects of charge conjugation (C), parity reversal (P), and time reversal (T) on the recorded data of an experiment involving elementary particles yields the same results as directly viewing the experiment.

Dark energy: In astrophysics and cosmology, a form of energy that permeates all of space and explains the recently observed acceleration in the rate of expansion of the universe.

Dark matter: Matter, not yet seen in laboratory experiments, that is distinct from ordinary matter and would explain the observed lifetimes of galaxies and the rates of rotation of stars in galaxies.

Deep Inelastic Scattering (DIS) Experiments: Name given to experiments that probed the interior of hadronic matter during the late 1960s to provide the first experimental confirmation that quarks exist inside hadronic matter.

Delbruck scattering: An experimentally verified process that demonstrates, due to quantum mechanical effects, that photons scatter off other photons.

Diffraction: The ability of a wave to bend around corners. This phenomenon is easily seen with photons (or light), using an experimental apparatus to demonstrate regions of brightness and shadows. However, all waves demonstrate this property. It is the reason a person can be heard around a corner.

DNA (Deoxyribonucleic acid): A nucleic acid that contains the genetic instructions for all cellular life forms on Earth.

Dual resonance model: The original name of string theory.

e: Transcendental number, similar to pi, with the property that it allows the simplest way to calculate logarithms. Its approximate value is 2.71828. The *e* stands for the first letter in the last name of Leonhard Euler.

E_8: Refers to one of the *generalized rotation groups* found by the mathematician Cartan. He showed that all possible mathematical descriptions of rotation-like symmetries form four infinite families and six "exceptional" ones. The *E* in E_8 stands for "exceptional."

Einstein's hypotenuse: A term created in this course to emphasize the important fact that Einstein's special relativity theory can be viewed as a definition of a less well-known geometry—one distinct from that commonly encountered in a non-technical education. For this course, this concept is a critical foundation used to gain more than a casual understanding of superstring/M-theory specifically and modern physics more generally.

Electromagnetic field: The field of forces (associated with electric charges and magnets) that possesses electric components, magnetic components, and a definite quantity of electromagnetic energy.

Electromagnetic interaction: One of the four forces in nature to which all electrically charged elementary particles are subject. The electromagnetic interaction possesses a single quantum known as the *photon*. When this force is studied only as acting on electrons, the results are summarized as quantum electrodynamics (QED).

Electron: The most familiar of the subatomic particles in the lepton family and the first known elementary particle. It possesses a negative charge, as well as another kind of charge called *weak charge*. The electron is a stable particle.

Entropy: A mathematical way to measure order or disorder in systems with statistical behavior. A state of maximum disorder possesses the lowest entropy.

Escape velocity: The initial velocity required to send an object permanently away from a body possessing a gravitational force.

Event horizon: A sphere that surrounds the singularity, which possesses several distinct properties. Nothing within the event horizon is ever able to escape eventually being "sucked" into the singularity. In principle, a light ray can orbit along the surface of the singularity. Outside the event horizon, the normal laws of gravitation apply. In particular, objects outside the event horizon can avoid being consumed by the black hole singularity.

Exchange force: The force exerted on elementary particles that are fermions and only due to the fact that identical fermions must obey the Pauli exclusion principle. All other fundamental forces depend on the existence of conserved quantities, i.e., charge, energy, etc.

Exchanger: A term introduced simply in the context of this course. More formally, in both mathematics and physics literature, this concept is called by the name *generator*.

Extra dimensions: Described by the 18th-century scientist James Joseph Sylvester as "inconceivable," this concept suggests that the universe may possess more than the three directions associated with length, breadth, and thickness. These are also called *hidden dimensions*.

Fermions: Elementary particles that have a spin rate that is 1/2 times an odd integer times h-bar.

Feynman diagram: A graphical representation of the interaction between elementary particles and/or quanta of energy. Invented by Richard Feynman, these diagrams provide an extremely powerful tool for conceptualization and calculation.

Feynman Rules: Various elaborations of rules invented by Richard Feynman that allow the use of pictographic representations of quantum processes to be converted unambiguously into mathematical expressions. The expressions can then be compared with data from experimental observation.

Forms: P-forms are mathematical quantities describing the analogs of photons; they couple, not to charged particles, but instead, to charged branes.

Free fermions: This term is used in an approach to constructing genuine four-dimensional strings via the use of spinors that describe only supersymmetrical quantum theories in the presence of an anomaly. This construction is rather unique in that, here, the anomaly does not destroy the mathematical consistency of the constructions. These fermions are added to a four-dimensional bosonic string to provide a complete description.

Frequency (light): Technically describes the rapidity with which either the electric or magnetic waves associated with a beam of light complete their oscillations when viewed from a point fixed in space. The frequency of a light beam is found by dividing the speed of light by the beam's measured wavelength.

Fudge factor: A value or parameter that is introduced into a mathematical formulation in an ad hoc way so as to produce a desired result.

Gauge invariance: The property that, though two or more photons might be different, if the manner in which they change in space and time is the same, then the electromagnetic fields the distinct photons produce are identical. More generally, this holds for any massless, force-carrying, spin-1 boson, and there are extensions to other values of spin.

General relativity: Albert Einstein's theory of gravitation developed by extension of the theory of special relativity to accommodate accelerating reference frames. It contains the introduction of the equivalence principle, which posits that gravity and acceleration are indistinguishable from each other. (This work is widely known as the theory of general relativity, or GR theory.)

Geometrical point particle: The basic mental construct that enables all of classical physics as enunciated by Isaac Newton. The idea is to imagine an object of no measurable size, which nonetheless has the mass of ordinary objects.

Geometry: The mathematical study of the rules and properties of figures, angles, and similar constructs.

Glashow/Salam/Weinberg model: The use of the mechanism involving the Goldstone particle to create appropriate masses for the W-plus, W-minus, and Z-zero quanta of the weak interaction. The mathematics of this demands the presence of a Higgs particle to generate mass for all other elementary particles.

Gluon(s): These are to the strong interaction as the photon is to the electromagnetic interaction. These eight bosons are the quanta of this force.

G minus 2 experiment: A series of different experiments, almost continuously run for more than 40 years, designed to study whether an electrical property for elementary particles predicted by quantum theory is experimentally verified. This is the most accurately tested scientific observation ever made.

Goldstino: A fermionic analog of the Goldstone particle, whose role is to give mass to the gravitino, the spin-3/2 superpartner to the graviton.

Goldstone particle: A bosonic particle that potentially has the mass of a tachyon but avoids this fate by "rolling down a potential energy hill." In the process, this particle can cause force-carrying, spin-1 bosons to acquire mass.

GPS (Global Positioning System): A satellite navigation system, originally created by the U.S. military, to allow the accurate determination of position on the surface of the Earth.

Gravitation interaction: One of the four forces in nature to which all elementary particles (including gravity's own quantum, the graviton) are subject. The word *gravity* is also used to describe this force.

Gravitino: The superpartner to the graviton.

Graviton: The force of gravity or gravitational interaction in the same way that the photon is the electromagnetic interaction. This is the quantum of the gravitational force.

Grey-body factors: When light falls on an object in the real world, some is typically reflected and some is absorbed. A body that totally absorbs light is called a *black body*, and a body that totally reflects light is a *white body*. A *grey body* is one in which the light is mostly absorbed. Typically also, the amount of reflection versus absorption depends on the color (that is, the wavelength) of the light. Thus, a *grey-body factor* is a mathematical description of how the relative amount of reflection versus absorption depends on the color.

Hadrons: Subatomic particles constructed as a composite of quarks, anti-quarks, and gluons.

h-bar: A physical constant of nature that plays a fundamental role in quantum theory. The *h* by itself is called *Planck's constant*, named for its discoverer, Max Planck, and h-bar is this number (h) divided by 2 × pi. Sometimes, h-bar is called the *Dirac constant* after the physicist Paul Dirac or, alternatively, the reduced Planck's constant.

Higgsino: A superpartner to the Higgs particle. Because Higgs particles are bosons, Higgsinos are fermions of spin rate 1/2.

Higgs particle(s): In the standard model of elementary particles, the only known consistent description of the massive force carriers of the weak interaction (the Z-zero, W-plus, and W-minus bosons) requires the use of an elaborate procedure known as the *spontaneous symmetry-breaking mechanism*. In this mechanism, a Goldstone particle (or more than one Goldstone particle) "rolls" down an energy hill to allow the force carriers to gain mass. In all known realizations of this idea, there remains one spin-0 boson (or possibly more) with mass that allows all fermions to acquire mass.

Hypotenuse: In a right triangle, refers to the side that is opposite to the right angle, that is, the angle of 90 degrees.

Intermediate vector bosons (IVB): Collective name for the Z-zero, W-plus, and W-minus bosons. These are to the weak interaction as the photon is to the electromagnetic interaction. They are the quanta of this force.

Invariance: The property of remaining unchanged by the action of a symmetry or some other definition involving change.

K-theory: A branch of the mathematics combining ideas from algebra and geometry that allows the study of whether the mathematical analogs of knots or tears occur on surfaces. Clifford algebras, which are related to spinors, play a role in K-theory.

Large Hadron Collider (LHC): Refers to a particle collider expected to be completed in 2007. The LHC is located in Geneva, Switzerland, and is to be operated at the European Center for Nuclear Research (known by its French acronym, CERN). The purpose of this device is to carry out research to understand the quantum world of nature in the greatest possible detail. If the Higgs boson or superpartners exist, they will most likely be discovered at LHC.

Laser Interferometer Gravitational-Wave Observatory (LIGO): A project designed to make the first direct experimental observation of waves of gravity, as predicted by general relativity theory of 1916.

Leptons: A family of six elementary particles—distinct from gauge bosons and quarks—that each possesses a spin of 1/2; leptons do *not* experience the chromodynamic force but are subject to the electroweak force and the gravitational force.

Lightest supersymmetric particle (LSP): In a universe where supersymmetry is valid, among denizens of the quantum world, the superpartners must occur with mass scales. Though these scales may be several hundred to a thousand times as massive as ordinary matter, one superpartner is the lightest. This is the LSP. In many models, the LSP is an excellent candidate for dark matter.

Loop quantum gravity (LQG): A proposed alternative to string theory to create a consistent theory for quantum gravity. The starting point of LQG is the theory of general relativity without modifications. An important role is played by a mathematical quantity called a *plaquette*. As of this date, attempts to use LQG to reproduce the Bekenstein-Hawking formula describing the relation of entropy for a black hole have been problematic. Even if this issue is resolved, unlike string theory, LQG has shown no signs of being a unified field theory.

Lorentz transformation(s): A set of equations introduced by Henrik Lorentz prior to the 1905 advent of relativity. Einstein adopted these to explain how measurements of space and time in one frame of reference are related to those in a second frame moving at a constant speed with respect to the first.

Metric: A mathematical quantity that specifies how lengths, areas, volumes, and other measures are to be determined. This is the basic quantity that controls the rules of geometry. In physics, this quantity determines gravitational forces, along with the growth and evolution of the universe.

Mode(s): In an extended system undergoing repeated oscillatory motion, overall recurring patterns may occur. These patterns may then be described as the *modes* of the vibration.

M-parton: The hypothetical point-like structures that may lie at the heart of M-theory. If M-theory is ever definitively demonstrated, it is likely that the strings of string theory will resemble a structure more like a string of pearls than a continuous pasta-like filament.

M-theory: An overarching mathematical structure within which all consistent string theories, as well as 11-dimensional supergravity, seem to emerge as special limits. No complete definitive proofs for the existence of this structure are known, but there are many pieces of circumstantial evidence.

Muon or mu particle: One of the particles in the lepton family. It possesses all the same properties as the electron except it is 105 times as massive.

Neutrino(s): Refers to any one of three particles in the lepton family. Neutrinos possess the same properties as the electron except they are electrically neutral and are thousands of times less massive.

Neutron: One of the two most familiar of the subatomic particles in the baryon family. The neutron possesses a positive electrical charge and is 1,839 times more massive than the electron. The neutron decays typically after 1,013 seconds as a free particle. It is usually found in the nucleus of atoms.

Neutron star(s): Star or stars of super-dense neutron matter that possess a powerful gravitational attraction, so that only neutrinos and x-rays can escape the interior of such stars.

Newton's law of universal gravitation: Mathematical result proposed by Isaac Newton to describe the gravitational force between planetary or stellar bodies.

Node(s): In an extended system undergoing repeated oscillatory motion, points may occur that are at rest; these are described as the *nodes* of the vibration.

Noether's theorem: States that whenever there occurs a symmetry that may change in space and time (that is, a gauge symmetry), there must be an associated conserved charge accompanied by its own current.

Non-Euclidean geometry: Refers to situations in which properties of figures, angles, and similar constructs are totally different from the (Euclidean) geometry typically taught in pre-college education.

Numeracy: The analog to literacy in the manipulation of numerical and mathematical data.

Parity reversal: The act of recording an experiment involving elementary particles by taking mirror-image data from the experiment.

Particle physics: The branch of physics (also called high-energy physics) that studies the elementary particles associated with energy and matter and their interactions.

Particle zoo: A term used colloquially to describe the extensive list of known elementary particles (they almost look like the list of animals in a large zoo).

Pauli exclusion principle: No two identical fermions can simultaneously possess the exact same quantum numbers.

Photino: The superpartner to the photon. Because the photon is a boson, the photino is a fermion of spin rate 1/2.

Photon: The quanta of the electromagnetic force, also called a particle of light.

Pi: The number obtained by dividing the circumference of a circle by its diameter. Its approximate value is 3.14159.

Planck energy: The amount of energy found by multiplying the fifth power of the speed of light by h-bar, next dividing this by Newton's constant, and finally, calculating the square root of this answer.

Planck length: The length that is found by first multiplying h-bar by the Newton universal gravitational constant, next dividing this result by the third power of the speed of light, and finally, taking the square root of the complete calculation.

Polarization (light): The property of light that describes the pattern swept out by either the electric or magnetic waves associated with a light beam as it passes a single point fixed in space.

Polarization (circular): If the polarization pattern swept out by the electric and magnetic waves associated with a light beam forms a circle, the light is *circularly polarized*.

Polarization (linear): If the polarization pattern swept out by the electric and magnetic waves associated with a light beam forms a line, the light is *linearly polarized*.

Positron: The antiparticle to the electron.

Power of 10: A phrase used to indicate either a multiplication or division by a factor of 10. For example, $100 = 10 \times 10$, so 100 is 10 to the power of 2; dividing by 100 is 10 to the power of -2. Similarly, $1000 = 10 \times 10 \times 10$, so 1000 is 10 to the power of 3.

Principia: *Philosophiae Naturalis Principia Mathematica* (Latin for "mathematical principles of natural philosophy"), a three-volume work by Isaac Newton, published in 1687, that established the foundation of physics.

Proton: One of the two most familiar of the subatomic particles in the baryon family. It possesses a positive electrical charge and is 1,836 times more massive than the electron. The proton is also a stable particle and is usually found in the nucleus of atoms.

Pulsar(s): Rotating neutron stars possessing strong magnetic fields that can eject high-energy radiation from their north and south magnetic poles. The rotation causes the poles to point in constantly varying directions so that, when viewed from a fixed position, they seem to pulsate.

Pythagorean Theorem: One of the most famous results of geometry, as described by "the square of the hypotenuse of a right triangle is equal to the sum of the squares of its sides."

Quantum, quanta: The smallest discrete part(s) of energy or matter in the view of quantum theory.

Quantization: This word has a number of meanings in physics. One definition, the one most commonly used in this course, refers to the fact that energy associated with any of the four fundamental forces comes in discrete packets, not in a continuum.

Quantum computer: Theoretical type of computer in which the computations using bits (the basic unit of information in a computer) are performed by manipulating "entangled" bosons (such as the photon) or fermions instead of electrons, as is done in present-day computers.

Quantum entanglement: Quantum theory suggests that objects that appear to be distinct in everyday observation are, in fact, associated with a type of wave, which contains information on all possible measurable attributes of the objects. The waves associated with distinct objects, under some circumstances, can be made to profoundly overlap. When this occurs, the information about the distinct objects becomes "'entangled" so that a subsequent measurement of a property of one object contains information about the other(s). Photons are examples in which it is possible to investigate scientifically this phenomenon.

Quantum numbers: The mathematical description of all the completely observable properties of an elementary particle.

Quantum theory: One of the two foundational theories of 20th-century physics, postulating that matter and energy are capable of manifesting behaviors similar to particles or waves, depending on the type of observation to which they are subjected.

Quantum tunneling: The process by which an object, typically an elementary particle, is able to travel through a region with a large energy barrier even though the object does not possess enough energy to accomplish this using the laws of classical physics (everyday experience). In the everyday world, a ball must be thrown upward very fast to travel very high. Via quantum tunneling, even if an electron were "thrown" upward very slowly, it would still be capable of reaching a very great height.

Quarks: A family of 18 elementary particles, each possessing a spin rate of 1/2 and subject to the effects of the chromodynamic force, the electroweak force, and the gravitational force. To date, quarks have been observed only in hadronic matter, occurring in triplets (baryons) or matter-antimatter pairs (mesons).

Regge trajectories: Graphical plots in which two mutually orthogonal axes are labeled by the magnitude of the spin rate and the square mass. Tulio Regge noticed that, in many reactions involving the production of new hadrons, a subset of the new hadrons possessed properties remarkably similar to the ones that were present before the collisions. When the mass and spin rate of this subset are placed on such a plot, they fall along straight lines, which are called *Regge trajectories*.

Riemann curvature tensor: A mathematical quantity, related to the metric, constructed such that its values will be the same for two observers whose different descriptions of geometrical properties are due only to their different choices of reference frame.

Scalar: Real and mathematical objects that possess the following property: After an observer walks around them for any distance along a circular path, they are found to look the same as they did on the initial viewing.

Selectron: The superpartner to the electron. Because the electron is a fermion, the selectron is a boson of spin rate 0.

Singularity: A geometrical point possessing no size at all that nonetheless has a mass that ranges from about five times that of the Sun up to virtually infinity. A singularity is the essential heart of a black hole.

Soliton(s): Stands for "solitary wave" and is applied to their mathematical description and the actual waves seen in nature. Solitons many consist of a few non-repeating waves, and their mathematical descriptions have the properties that these waves "bounce" off one another like physical objects.

Special relativity: One of the two foundational theories of 20th-century physics, this theory suggests that the speed of light is measured to be the same for all observers in motion without acceleration and consistently relates the measurements performed by such observers.

Speed of light: The speed at which light travels in a vacuum, accurately represented by 299,792,458 meters per second. According to Einstein's work, no object or process containing information can travel faster than this rate, and only massless objects can achieve it.

Spin: The intrinsic angular momentum of an elementary particle, which is the property of appearing to have attributes of tiny spinning balls, even though they possess no size at all. The rates of spin of elementary particles are measured in terms of a quantity called *h-bar* and come in any integer or half-integer multiplied times this rate.

Spinning string: A string theory that incorporates the mathematics used to describe the spin of elementary particles. Its vibratory modes do not automatically incorporate equal numbers of bosonic and fermionic "notes" but can be made to do so with a certain truncation.

Spinning string ($N = 1$): A string that incorporates the mathematics of spin as described for elementary particles. The key point of the construction is to introduce one additional anti-commuting strand for each real strand of the string. However, the anti-commuting strands behave as vectors when viewed before and after a complete circuit is made about them.

Spinning string ($N = 2$): A string that incorporates the mathematics of spin as described for elementary particles. The key point of the construction is to introduce one additional anti-commuting complex strand (that is, using the square root of -1) for each complex strand of the string. Once again, the anti-commuting strands behave as vectors when viewed before and after a complete circuit is made about them.

Spinor: Real and mathematical objects similar to arrows (or vectors) that possess the following property: After an observer walks around them in a complete circle any *odd* number of times, they are found to point in the *opposite* direction as that seen in the initial viewing. After an observer walks around them in a complete circle any *even* number of times, they are found to point in the *same* direction as that seen in the initial viewing.

Spontaneous symmetry breaking: When the Goldstone particle is at the "top of the energy hill," all directions down the hill are equivalent. This is the meaning of a *symmetry*. After the Goldstone particle rolls down the hill, the direction it takes becomes special and the symmetry is broken.

Squark: A superpartner to any of the quarks. Because quarks are fermions, squarks are bosons of spin rate 0.

Stanford Linear Accelerator Center (SLAC): A national laboratory in the United States operated by Stanford University. The 1.9-mile-long (3-kilometer) straight accelerator is underground and was the site of the DIS experiments that established the existence of quarks.

Stern-Gerlach experiment: The 1921 experiment conducted by Otto Stern and Walter Gerlach showing that electrons behave as spinors.

String theory: A mathematical theory proposed for describing all of physical reality; it suggests that modes of vibration of almost infinitesimal strings correspond to measurable fundamental properties of energy, matter, space, and time.

Strong interaction: One of the four forces in nature to which only quarks are directly subject (it indirectly acts on all hadrons because quarks occur in the interior of hadronic matter). The strong force possesses eight quanta known as *gluons*. This force is sometimes also called *quantum chromodynamics* (QCD).

Superconducting Super Collider (SSC): A particle accelerator designed (but never constructed) to investigate the quantum world of nature and determine which of its possible mathematical descriptions actually occurs. Such a facility would be required to determine whether nature contains one or more Higgs particles, superpartners, and so on, or if some totally unexpected feature is present.

Supergravity: An extension of the theory of general relativity in which at least one fermionic superpartner of spin rate 3/2 is consistently included in the mathematics of general relativity theory.

Super-Kamiokande (or Super-K): A neutrino physics laboratory designed primarily to study flux of neutrinos from the Sun, from supernovae, and in the atmosphere. Super-K was the site where the discovery of neutrino masses was confirmed.

Superparticle: A mathematical description of both a particle of ordinary matter and/or energy simultaneous with its superpartner.

Superpartner: In most systems possessing supersymmetry, for each boson, there must be a fermion that can be continuously "mixed" with it. The bosons and fermions allowed to mix this way are *superpartners* to one another.

Superspace: A mathematical construct in which some directions are described by ordinary numbers with the usual commuting multiplication laws and some directions are described by Grassmann numbers with anti-commuting multiplication laws.

Superstring: Superstring theory is a shorthand term for *supersymmetrical string theory* because, unlike bosonic string theory, it is a version of string theory that automatically incorporates equal numbers of bosonic and fermionic "notes" among its vibrations. The key point of this construction is to introduce additional anti-commuting strands that are multiples of the number 16 in addition to the 10 strands of a bosonic string. The anti-commuting strands behave as spinors when viewed before and after a complete circuit is made around them.

Superstring (heterotic): A superstring that is constructed by adding to the usual bosonic strands of the string 10 additional Grassmann strands with properties corresponding to a vector and an additional right-mover to describe symmetries related to the gauge sector. This string can only be closed.

Superstring (type-I): A superstring that is constructed by adding to the usual bosonic strands of the string 16 additional Grassmann strands with properties corresponding to one left-handed spinor with regard to the 10 dimensions of the string. The string can be either open or closed.

Superstring (type-IIA): A superstring that is constructed by adding to the usual bosonic strands of the string 32 additional Grassmann strands with the properties corresponding to one left-handed spinor and one right-handed spinor with regard to the 10 dimensions of the string. The string can only be closed.

Superstring (type-IIB): A superstring that is constructed by adding to the usual bosonic strands of the string 32 additional Grassmann strands with the properties corresponding to two left-handed spinors with regard to the 10 dimensions of the string. The string can only be closed.

Supersymmetry: A mathematical symmetry between bosons and fermions in which there is no way to determine how much energy in a system is due solely to the bosons separately or solely to the fermions separately.

Surface of last scattering: For a time after the Big Bang, the density of matter in the universe was so high that light could not freely travel through it. Recent results from the Relativistic Heavy Ion Collider (RHIC) suggest that the universe may have been rather similar to a milky liquid during this phase. As the universe continued to expand, an instant occurred when photons, particles of light, were first freed to travel. In the instant before this, they had one last "bounce" from the "surface of last scattering," a process rather like setting off a flashbulb. This "snapshot" of the infant universe can be seen today using microwaves (see **cosmic microwave background**).

SUSY: The abbreviation for "supersymmetry" or "supersymmetric system," in which certain correspondences occur between bosons and fermions.

Symmetry: An action in physics or mathematics that, when applied to some object, real or mathematical (a number, function, etc.), has the effect that before and after implementation of the action, the object looks the same.

Symmetry (continuous): A symmetry that can be applied only in a continuous manner. As an example, imagine a round poker chip laying on a table and a game with two players. The game begins with the two in the room looking at the chip. Player A is sent out of the room, and player B may choose to rotate the chip on the table through any size of angle. When player A returns to the room, there is no way to decide through how much angle the chip was turned or whether it was turned at all while player A was outside the room.

Symmetry (discrete): A symmetry that can be applied only in discrete "jumps." As an example, imagine a poker chip laying on a table and a game with two players. The game begins with the two in the room looking at the chip. Player A is sent out of the room, and player B may choose to flip over the chip or not. When player A returns to the room, there is no way to decide which action took place while he or she was outside the room.

Tachyon: A type of never-observed elementary particle that, of necessity, travels no less than the speed of light. Tachyons also gain energy as their velocity is lowered, precisely the opposite of ordinary matter or elementary particles.

Taon or tau particle: One of the particles in the lepton family. It possesses all the same properties as the electron, except it is 1,700 times as massive.

Time-like particle(s): A particle that possesses negative energy due only to its motion and appears typically when the gauge invariances associated with massless, force-carrying, spin-1 bosons are disturbed.

Time reversal: The act of recording an experiment involving elementary particles by first filming a movie of it, then taking data by viewing the movie while it is run in reverse.

Twistors: The analogues of spinors for spaces in which there is more than one direction that appears in Einstein's hypotenuse with the same sign as the usual "time" direction.

Unified field theory: A mathematical construct that contains all the elements to describe energy, matter, space, and time and gives an accurate and consistent view of our universe. Einstein (who labored unsuccessfully at its construction) suggested this name. Currently, this is also called by the immodest and misleading name, a *theory of everything*.

Vacuum: The state of lowest energy in a physical system. In quantum theory, the vacuum is not the state of emptiness. Instead, via quantum processes, antiparticle/particle pairs spontaneously come into existence in the vacuum and annihilate each other.

Vacuum polarization: A quantum process whereby a force-carrying, spin-1 boson has part of its energy converted into any number of antiparticle/particle pairs that eventually annihilate each other, thus returning all their energy back to the original boson.

Vector: Real and mathematical objects similar to arrows that possess the following property: After an observer walks around them in a complete circle any number of times, they are found to point in the same direction as that seen in the initial viewing.

Wave function: In quantum theory, a mathematical function used to describe the propagation of the wavelike properties associated with any particle or group of particles.

Wavelength (light): Technically describes the crest-to-crest distance of either the electric or magnetic waves associated with a light beam. To the human eye, this is associated with the color of the light.

Weak interaction: One of the four forces in nature to which all elementary particles (with the exception of the graviton) are subject. It possesses three quanta known as the *W-plus*, *W-minus*, and *Z-zero* bosons. This force is sometimes also called *quantum flavor dynamics* (QFD).

Yang-Mills equations: Equations developed by C. N. Yang and R. Mills that describe the physics of gluons and IVBs. These equations are the analogs of those developed by Maxwell.

Zero-point energy: A classical oscillator, such as the pendulum of a clock, can hang undisturbed at rest. The laws of quantum theory forbid a pendulum in the quantum world from doing this. Because a quantum pendulum must always possess some motion, it must always have an associated energy. The smallest value of this is the zero-point energy.

Biographical Notes

Banks, Thomas (born 1947): American physicist who, in 1999, in collaboration with Fischler, Shenker, and Susskind, gave the strongest evidence to date for the existence of M-theory at a fundamental level.

Bardeen, John (1908–1987): American physicist and the only person to be awarded the Nobel Prize in Physics…twice—first with Shockley and Brattain, in 1956, for the development of the transistor and then with Cooper and Shrieffer, in 1972, for achieving an understanding of superconductivity.

Bekenstein, Jacob (born 1947): Professor of theoretical physics at the Hebrew University of Jerusalem who contributed significantly to the science of black- hole thermodynamics.

Bell, John (1928–1990): Irish mathematical physicist who contributed to the deeper understanding of the foundation of quantum theory through his work known today as *Bell's inequality*.

Bohr, Niels (1885–1962): Danish physicist; with Einstein, the leading architect of the "old" (1905–1925) quantum theory of matter and energy and the leading interpreter of quantum mechanics in its formative years (1925–1935).

Cartan, Élie (1869–1951): French mathematician who studied rotations, among other topics, and wrote the first mathematical description of spinors.

Clifford, William (1845–1879): English mathematician who studied non-Euclidian geometry and developed what came to be known as *Clifford algebra*. His ideas played a fundamental role in Einstein's theory of general relativity.

Cremmer, Eugene (born 1942): Working with Julia and Scherk in 1978, he constructed the 11-dimensional supergravity theory that has led to the determined efforts to unravel M-theory.

De Broglie, Louis (1892–1987): French physicist whose 1923 doctoral dissertation extended early quantum theory by arguing that, just as waves behave like particles, particles must behave like waves. His prediction was confirmed three years later.

Deser, Stanley (born 1927): Professor of physics at Brandeis University; his research involves the study of quantum fields and gravitational theories.

Dirac, Paul (1902–1984): British electrical-engineer-turned-physicist, creator of the relativistic theory of the electron, and the founder of what became quantum electrodynamics. His prediction of the positron was a startling instance of the fertility of abstract mathematics for revealing physical reality.

Eddington, Arthur (1882–1944): Possibly the most important astrophysicist of the early 20th century, he headed joint expeditions to Sobral, Brazil, and the island of Principe to verify a prediction of Einstein's gravitational theory.

Einstein, Albert (1879–1955): German physicist who made several fundamental contributions to science. In 1905 alone, Einstein demonstrated the atomic origin of Brownian motion, provided compelling evidence for the quantum theory of matter, and produced the first installment of his theory of relativity. In later life, Einstein struggled to find a grander "unified theory," with no success.

Feynman, Richard (1918–1988): American physicist who, with Schwinger and Tomonaga, founded quantum electrodynamics. He invented the *Feynman diagram*, a simple sketch that allows one to think qualitatively about complex physical processes while retaining the mathematics "underneath" the pictures.

Fourier, Jean (1768–1830): French mathematician and physicist who is best known for a mathematical result known as the *Fourier series*, which provides an important tool to analyze the motion of any system undergoing repetitive motion.

Gauss, Johann (1777–1855): German mathematician who, in 1798, made one of the most important advances in geometry since the time of the Greeks; one of his discoveries is a foundation for the work of Maxwell.

Gell-Mann, Murray (born 1929): American physicist who formulated the *Eightfold Way*, organizing elementary particles into families and, with Zweig, created the quark theory of matter, later called *quantum chromodynamics*.

Gerlach, Walther (1889–1979): German physicist who, with Stern, conducted the Stern-Gerlach experiment, discovering the property of spin quantization.

Goldstone, Jeffrey (born 1933): American physicist who proved that, whenever symmetry is spontaneously broken, a particle called the Goldstone particle appears that can be used to give mass to spin-1 force carriers.

Grassmann, Hermann (1809–1877): German mathematician and linguist who invented a new algebra of vectors with applications in theoretical physics.

Green, Michael (born 1946): British physicist who, working with Schwarz, first formulated the superstring and later proved that all versions of the theory are free of anomalies.

Gross, David (born 1941): American physicist who, working with Harvey, Matrinec, and Rohm, invented the heterotic string in 1985.

Harvey, Jeffrey (born 1955): American physicist and co-discoverer of the heterotic string.

Hawking, Stephen (born 1942): British physicist, best known as a theoretician of space, time, gravity, and black holes. His work, along with that of Bekenstein, is fundamental to the search for a quantum theory of gravity.

Heisenberg, Werner (1901–1976): German physicist who developed the *uncertainty principle*, which states that one cannot, in principle, have precise simultaneous knowledge of the momentum and position of a particle.

Hertz, Heinrich (1857–1894): German physicist who, in 1888, showed that the prediction of Maxwell's equation regarding the transmission of electromagnetic signals through space was correct.

Higgs, Peter W. (born 1929): Scottish physicist and discoverer of the Higgs boson, a hypothetical elementary particle, strongly expected to be discovered in experiments to be undertaken between 2007–2020.

Jackiw, Roman (born 1939): Polish-born American physicist who, working with Bell, discovered the essential nature of anomalies in quantum theory.

Julia, Bernard (born 1952): With Cremmer and Scherk, he constructed the 11-dimensional supergravity theory, leading to efforts to unravel M-theory.

Kaluza, Theodor (1885–1954): Succeeded, in 1921, at the task that had eluded both Maxwell and Einstein by, with the introduction of one extra dimension, finding a unified description of electromagnetism and gravitation.

Kemmer, Nicolas (1911–1998): English physicist who, in 1938, offered the proposal that the similarities between the neutron and proton could be interpreted simply by considering that their charges are related to angles that exist in isotopic charge space. Unlike the idea of Kulaza and Klein, Kemmer's does not require the existence of extra dimensions.

Klein, Oskar (1894–1977): Swedish-American physicist who introduced compactification, required to consider that a mathematical universe with extra dimensions is not automatically ruled out by experimental observation.

Lederman, Leon (born 1922): Former director of Fermilab and professor of physics at the Illinois Institute of Technology, who shared the Nobel Prize with Schwartz and Steinberger for work with neutrino beams and has written *The God Particle* and other books aimed at increasing public understanding of science.

Leibniz, Gottfried (1646–1716): German philosopher and mathematician credited, independently of Newton, with the invention of calculus.

Lobachevsky, Nikolai (1792–1856): Russian mathematician who, in 1829, become the first person to publish a work on non-Euclidean geometry.

Lorentz, Henrik (1792–1856): Dutch physicist and 1902 Nobel Prize recipient for his work exploring the relation of magnetism and radiation. Although the conceptual framework of his derivation of Lorentz transformations relied on assuming the existence of the ether, Einstein adopted these and showed their consistency without requiring its existence.

Maldecena, Juan (born 1968): Italian physicist best known for his discovery of the AdS/CFT correspondence (also known as the *Maldecena conjecture*).

Martinec, Emil, (born 1958): American physicist and co-discoverer of heterotic string theory.

Maxwell, James Clerk (1831–1879): Scottish physicist who presented four equations that codify every aspect of electromagnetism, including the previously unrecognized phenomenon of electromagnetic radiation. Maxwell's equations paved the way for the discovery of relativity and form the classical underpinnings of quantum electrodynamics, the quantum theory of light.

Mendeleev, Dmitri (1834–1907): Russian chemist who systematized the weights and chemical properties of 63 chemical elements in his periodic table of the elements.

Michell, John (1724–1793): British theologian, geologist, and astronomer who was the first person to consider the possibility of black holes and is given credit for proposing the existence of binary stars.

Minkowski, Hermann (1864–1909): Polish-German mathematician who, in 1907, realized that Einstein's theory of special relativity has a simple interpretation in terms of space-time geometry (*Minkowski space*). Without this realization, it is doubtful Einstein would have been able to create his 1916 opus on general relativity.

Neveu, Andre (born 1942): French physicist who, with Schwarz, is credited with evolving the second generation of string theory (spinning strings).

Newton, Isaac (1642–1726): British natural philosopher and mathematician who developed calculus, the laws of motion, the law of universal gravitation, and principles of optics and light. Many of his ideas were summarized in the *Principia* of 1686.

Noether, Emmy (1883–1935): German-born mathematician who published, in 1918, *Noether's theorem*, a work directly relevant to particle physicists.

Pauli, Wolfgang (1900–1958): Austrian physicist whose 1925 exclusion principle provided the first systematic explanation of the periodic table and opened the way to Schrödinger's and Heisenberg's versions of quantum mechanics. The Pauli exclusion principle states that no two quantum particles (of spin 1/2) may exist in the same quantum state. Pauli also predicted the existence of a new particle of nature, later called the *neutrino*.

Penzias, Arno (born 1933): German-born American physicist whose work (with Wilson) with a receiver called the *Big Ear*, monitoring radio emissions for the Milky Way's encircling gas ring, led to the detection of the cosmic microwave background.

Planck, Max (1858–1947): German physicist who, in 1900, proposed the idea that energy comes in discrete bundles, called *quanta*, at the atomic scale. This theory helped to explain the spectrum of electromagnetic radiation emitted by a "black body" that absorbs all electromagnetic radiation that falls upon it.

Ramond, Pierre (born 1943): American physicist who independently discovered spinning string theories.

Randall, Lisa (born 1962): American physicist whose work with Sundrum posits that with the acceptance of the existence of at least one extra dimension and that our universe exists in a manner similar to a pane of glass, there is a simple reason to understand why the force of gravity is so much weaker than other fundamental forces.

Riemann, Bernhard (1826–1866): German mathematician whose contributions to geometry paved the way for Einstein's development of the theory of general relativity.

Rohm, Ryan (born 1969): American physicist and co-discoverer of the heterotic string.

Russell, John (1808-1882): Scottish engineer who first reported the observation of a solitary wave, which he called the *wave of translation*, and made one of the first observations of the Doppler shift of sound frequency.

Salam, Abdus (1926–1996): Born and raised in Pakistan, he made major contributions in the mathematical development of the standard model, in particular, the unification of the weak and electromagnetic forces of nature.

Scherk, Joel (1947–1979): French physicist who, with John Schwarz, was responsible for the realization that string theory possessed a mathematical description of the graviton and was, thus, capable of realizing Einstein's final dream of a unified field theory. Tragically, he committed suicide in 1979.

Schrödinger, Erwin (1887–1961): Austrian physicist; one of the giants of quantum mechanics. His wave mechanics of 1925 became the basis of Dirac's relativistic theory of the electron, which evolved into quantum electrodynamics.

Schwarz, John (born 1941): American physicist who, with Joel Scherk and independently of Pierre Ramond, invented the spinning string.

Schwinger, Julian (1918–1994): American physicist who shared a Nobel Prize in Physics for the complete theory of quantum electrodynamics, the quantum theory describing the interactions between electrons and photons.

Siegel, Warren (born 1951): American physicist who provided one of the major tools (called *chiral bosons*) necessary for the construction of the heterotic string, among many other fundamental contributions to field theory and supersymmetry. Dr. Siegel and I developed a third formulation of the heterotic string that is directly connected to the gauge transformations of the standard model and opened the way to a description of a four-dimensional version of heterotic string theory.

Stern, Otto (1888–1969): German physicist; together with Walther Gerlach, he conducted the Stern-Gerlach experiment, through which the property of spin quantization was discovered.

Stoney, George (1826–1911): British physicist credited with first proposing and later naming the first elementary particle, the electron.

Sundrum, Raman (born 1962): Indian-born American physicist who, in collaboration with Randall, proposed the *brane world scenario* (also called Randall-Sundrum models) as a model for the physical universe, in one of the most highly cited research papers of the era.

Susskind, Leonard (born 1940): American physicist who first realized that 18th-century mathematics relevant to elementary particle physics could be interpreted as arising from interactions of minute, filament-like objects subject to the laws of special relativity.

t'Hooft, Gerardus (born 1946): Dutch physicist and 1999 Nobelist who, in collaboration with Veltman, developed a practical way to calculate the effects of quantum theory of electromagnetic and weak interactions.

Tomonaga, Sin-Itiro (1906–1979): Japanese physicist who shared a Nobel Prize in Physics (with Feynman and Schwinger in 1965) for the complete theory of quantum electrodynamics, the quantum theory describing the interactions between electrons and photons.

von Haidinger, Wilhelm (1795–1871): Austrian geologist and physicist who discovered that he could visually detect the polarization of light. This effect is now called *Haidinger's brush*.

Wigner, Eugene (1902–1995): Hungarian physicist and mathematician who received the Nobel Prize in Physics in 1963 "for his contributions to the theory of the atomic nucleus and the elementary particles, particularly through the discovery and application of fundamental symmetry principles."

Wilson, Robert (born 1936): American physicist whose work led to the detection of what is now known as the cosmic microwave background.

Witten, Edward (born 1951): Mathematical physicist at the Institute for Advanced Study in Princeton who has made a number of contributions to the field of string theory and is the founder of M-theory.

Zumino, Bruno (born 1923): One of the leading experts in field theory and credited, working with Julius Wess, with constructing one of the first supersymmetric theories in four dimensions.

Zweig, George (born 1937): Originally trained as a particle physicist under Richard Feynman, he proposed the existence of quarks independently from Murray Gell-Mann. He later moved into the field of neurobiology.

Bibliography

Abbott, Edwin. *Flatland: A Romance of Many Dimensions*. New York: Dover Publications, 1884. This work is well known among scientists and nonscientists alike as a fanciful story about a two-dimensional world that nonetheless displays real mathematical sophistication wrapped about a parody of life in 19th-century England, a rigid society stratified by class.

Ashton, Anthony. *Harmonograph: A Visual Guide to the Mathematics of Music*. New York: Walker & Co., 2003. Animations in this course were sometimes used to represent sound visually, which in turn, allowed us to use the similarities between the mathematics of strings and that of sound to construct useful analogies. An actual device from the 19th century directly constructed visual images from sound. This book explains the device and contains images of sound obtained by using it.

Bartusiak, Maria. *Einstein's Unfinished Symphony*. New York: Berkley Books, 2003. This book explores the context, history, and developments in the search to detect the waves of gravity predicted by Einstein's 1916 work. Such waves have never been detected, but Bartusiak presents a fun and informative introduction to the subject based on more than a decade spent following such developments.

Bernstein, Jeremy. *Einstein and the Frontiers of Physics*. Oxford: Oxford University Press, 1996. This book contains biographical information about Einstein, records of his interactions with other scientists, and discussions of the areas in physics that were paramount in his mind.

Burger, Edward B., and Michael Starbird. *The Heart of Mathematics*. Emeryville, CA: Key College Publishing, 2000. This book contains introductory information on many areas of mathematics, such as number theory, topology, knot theory, and geometry, that are critical to obtain a truly well-informed view of string theory.

Cajori, Florian. *A History of Mathematics*, 5th ed. New York: Chelsea Publishing Co., 1991. This book tells the story of progress in mathematics from the period of antiquity to the end of the First World War. Special attention is paid to the 19th and early 20th centuries.

Cheetham, Nicolas. *Universe: A Journey from Earth to the Edge of the Cosmos*. London: Smith-Davies Publishing, 2005. A beautiful collection of images highlights this book, illustrating many cosmological structures, including black holes, nebulae, planets, stars, supernovae, and white dwarfs.

Close, Frank, Michael Marten, and Christine Sutton. *The Particle Explosion*. Oxford: Oxford University Press, reprinted 1994. This book does for the microcosm what Nicholas Cheetham's does for cosmology. It is a brilliantly illustrated book that includes chapters with imaginative imagery of the denizens of the quantum world interspersed with imagery of the instrumentation and devices that allow humanity to study the microcosm.

Darrigol, O. "The Spirited Horse, the Engineer, and the Mathematician: Water Waves in Nineteenth-Century Hydrodynamics," in William Dunham, *The Mathematical Universe*. Hoboken, NJ: John Wiley & Sons, 1997. An essay on the career and interests of John Scott Russell, notable for his description of what he called the "Great Wave of Translation," whose mathematical description would prove critical to the discovery of the heterotic string a century later.

Gates, Jr., S. James, "On the Universality of Creativity in the Liberal Arts and in the Sciences," in *Beyond Two Cultures: The Sciences as Liberal Arts*. Santa Barbara, CA: Institute for the Liberal Arts at Westmont, 2006. (http://www.westmont.edu/institute/pages/2005_program/panelists_2 005.html.) This essay contrasts fields in the sciences with those in the liberal arts and gives an extended description of my perception of the life and work of the theoretical physicist.

Gell-Mann, Murray. *The Quark and the Jaguar: Adventures in the Simple and the Complex*. New York: Henry Holt & Co., 1994. In this panoramic book, the Nobel Laureate physicist, who was also one of the scientists whose work was definitive in establishing the existence and description of quarks, sweeps through the breadth of fundamental theoretical physics.

Goldsmith, Donald. *Einstein's Greatest Blunder? The Cosmological Constant and Other Fudge Factors in the Physics of the Universe*. Cambridge, MA: Harvard University Press, 1995. In spite of its title, this book is actually a review of all of cosmology written in an accessible way. It touches on all the major issues in the field and gives explanations and an account of modern cosmology.

Greene, Brian. *The Elegant Universe*. New York: Vintage Books, 2000. The much-celebrated book that brought the topic of superstring/M-theory to the attention of the wider world and provided the basis for the NOVA/PBS television video program. The author shows a keen ability to guide his readers through a large part of the history and development of the subject.

———. *The Fabric of the Cosmos: Space, Time, and the Texture of Reality*, 1st ed. New York: Alfred Knopf, 2004. In this sequel to *The Elegant Universe*, Greene takes the reader deeply into the realm of questions about the very essence of space and time under the impact of discoveries in cosmology, particle physics, and superstring/M-theory.

Gribbin, John, and Mary Gribbin. *Annus Mirabilis: 1905, Albert Einstein, and the Theory of Relativity*. New York: Chamberlain Bros, 2005. In this work, the authors give a vivid description of Einstein's year of miracles, during which he announced to the physics community that a new "star" had appeared, although it would be another three years before this was apparent to the field.

Halpern, Paul. *The Great Beyond: Higher Dimensions, Parallel Universes and the Extraordinary Search for a Theory of Everything*. Hoboken, NJ: John Wiley & Sons, 2004. In this work, the author guides the reader through the maze of modern physics theory suggesting the universe may well possess extra dimensions that have not yet been detected through experimental instrumentation.

Harmon, P. M. *The Natural Philosophy of James Clerk Maxwell*. Cambridge: Cambridge University Press, 1982. All modern communication technology rests on Maxwell's work, which in this book, can be viewed through his scientific letters and papers. The work of Maxwell also provided the genesis for Einstein's miracle year with regard to special relativity.

Hawking, Stephen. *The Universe in a Nutshell*. New York: Bantam Books, 2001. In this work, the highly successful author-scientist carries the reader into the realms of black holes, dark matter, supergravity and supersymmetry, MACHOS, p-branes, proto-galaxies, and WIMPS. Also included are discussions of highly speculative prospects, such as time travel.

———. *A Brief History of Time*, 10th anniversary ed. New York: Bantam Books, 1998. This book marked an emergence, from the theoretical physics community into the popular realm, of the scientific discussion of the evolution of the universe. A challenging work that many regard as opening a new world for the public.

Hoddeson, Lillian, and Vicki Daitch. *True Genius: The Life and Science of John Bardeen*. Washington, DC: Joseph Henry Press, 2002. This book describes the life and work of the only individual in Nobel history to become the recipient of the Nobel Prize twice: the first for his contribution to the discovery of transistors and the second for his work in solving a mystery, that of superconductivity.

Kaku, Michio. *Hyperspace: A Scientific Odyssey through Parallel Universes, Time Warps, and the 10th Dimension*. New York: Anchor Books, 1995. This work marked perhaps the first widespread public discussion of deliberations that had been occurring within the physics community for more than a decade. The author-scientist is a popular writer, as well as a media presence on many television scientific documentaries, and offers the reader insight into his perspective on these developments.

Kane, Gordon L. *Supersymmetry: Unveiling the Ultimate Laws of Nature*. New York: Perseus Books, 2001. This book marked the first discussion of the possibility that the universe may possess a totally unexpected property that implies the existence of yet-to-be seen forms of matter and energy known as *superpartners*. If they exist, these would mirror known forms of matter and energy. Many readers have found this to be a most challenging book that requires a great deal of effort in order to gain its benefit.

Kanigel, Robert. *The Man Who Knew Infinity: A Life of the Genius Ramanujan*. New York: Simon & Schuster, 1991. This book follows the arc of the life and career of one of the most unusual mathematicians in human history, Srinivasa Ramanujan. Like Einstein, Ramanujan started his rise to prominence as a clerk, but his unusual work and accomplishments in the area of mathematics called number theory took him from India to the halls of Cambridge University and a collaboration with the English mathematician G. H. Hardy.

Kaplan, Robert, and Ellen Kaplan. *The Art of the Infinite: The Pleasures of Mathematics*. Oxford: Oxford University, 2000. Historical figures important to the development of ideas that today go almost unnoticed are revealed, along with diagrams and figures, in an effort to render a tough subject readable for the nonexpert.

Krauss, Lawrence M. *Hiding in the Mirror: The Mysterious Allure of Extra Dimensions, from Plato to String Theory and Beyond.* New York: Viking Books, 2005. The author-physicist is well known for his popular works on the physics of *Star Trek*. In this work, he turns his attention to the topic of extra dimensions and, in particular, guides the reader to a substantial history on its evolution through several millennia of human history.

Lederman, Leon. *The God Particle.* New York: Dell Publishing, 1994. The Nobel Laureate physicist-author of this work initiated the first large-scale public discussion of the theory that all mass in the universe possesses a single origin, the Higgs boson. Dr. Lederman provides a lively and accessible discussion in a voice that is unique among the physics community. The book also chronicles his career, which culminated in the recognition of the Nobel Prize for his work.

———, and C. T. Hill. *Symmetry and the Beautiful Universe.* Amherst, NY: Prometheus Books, 2004. In this work, Drs. Lederman and Hill undertake to open wide for public display the concept of symmetry. This concept has been central to the progress of fundamental physics since at least the time of Einstein. In fact, this was the primary guide that led him to his opus on general relativity.

Maor, Eli. *To Infinity and Beyond: A Cultural History of the Infinite.* Basel: Birkhäuser, 1987. One of the primary reasons behind the drive for a quantum theory of gravity is to avoid the presence of *infinities*. What are these? This fascinating book offers its readers insight into this question.

Miller, Arthur I. *Empire of the Stars: Obsession, Friendship, and Betrayal in the Quest for Black Holes.* Boston: Houghton Mifflin Co., 2005. Though Eddington was the scientist who announced to the world that Newton's view of gravity had been overthrown by observations that supported Einstein's, he was not receptive to the radical new ideas on black hole formation offered by Subrahmanyan Chandrasekhar. This book chronicles not just the conflict between these views but also the culture of the field during the early and mid-20th century.

Mlodinow, Leonard. *Euclid's Window: The Story of Geometry from Parallel Lines to Hyperspace.* New York: Free Press, 2002. Written by a young scientist who witnessed and was part of the period that brought superstring/M-theory to its current position in the field of physics, this book offers a necessarily personal view on a key member of the community that developed the theory. In addition, it opens a unique view for the reader on how geometry has evolved up to its current place in superstring/M-theory.

Motz, Lloyd and Jefferson Hane Weaver. *The Story of Physics*. New York: Avon Books, 1989. The authors of this book, one of a pair of very similar works, give a nice overview similar to the concept behind these lectures, describing the intricate interweaving of ideas and demonstrating linkages to concepts from antiquity to the present in the evolution of physics. The interplay between mathematics and physics is a highlight.

Motz, Lloyd and Jefferson Hane Weaver. *The Story of Mathematics*. New York: Plenum Press, 1993. The authors of this book, one of a pair of very similar works, give a nice overview similar to the concept behind these lectures, describing the intricate interweaving of ideas and demonstrating linkages to concepts from antiquity to the present in the evolution of mathematics.

Nahin, Paul J. *An Imaginary Tale: The Story of $\sqrt{-1}$*. Princeton, NJ: Princeton University Press, 1998. For most people, the concept that some numbers are more "complex" than others might seem odd. In fact, the numbers that are formally called *complex numbers* have had a 1500-year history to reach the status of an accepted idea. This book tells this story and reveals that the creation of mathematics resembles other human creative endeavors in surprising ways.

Oerter, Robert. *The Theory of Almost Everything: The Standard Model, the Unsung Triumph of Modern Physics*. Upper Saddle River, NJ: Pi Press, 2005. Dr. Oerter's work goes a long way toward pointing out for the public one of the towering scientific achievements of modern physics. The *standard model* is not speculative but one of the best tested pieces of science ever developed. This work lays out the intricacies of the scientific paradigm in a readable manner.

Penrose, Roger. *The Road to Reality: A Complete Guide to the Laws of the Universe*. New York: Alfred A. Knopf Publishing, 2006. One of the field's most distinguished members lays out arguments that suggest alternatives to the conventional wisdom of the field in this well-received work. The agenda of this book is not just to elucidate unusual physics but also to point out curious connections to the human consciousness in the perception of reality.

Randall, Lisa. *Warped Passages: Unraveling the Mysteries of the Universe's Hidden Dimensions.* New York: HarperCollins Publishers, 2005. This is a work by one of the most cited physicists in the world, who is also on the faculty at Harvard University. It has thus far shown all the signs of being a widely popular work, bringing physics to the public in an interesting and engaging way. The author presents the view held by a substantial part of the physics community that the extra dimensions of string theory are real. Perhaps an interesting twist is that it is not only superstring/M-theory that is simplified by this assumption.

Rucker, Rudy. *Infinity and the Mind: The Science and Philosophy of the Infinite.* Basel: Birkhäuser, 2004. This book, while providing commentary on the mathematical notion of infinity, also takes its readers to the realm of tantalizing philosophical questions.

Sagan, Carl. *Cosmos.* New York: Ballantine Books, reissued, 1997. Though dated and suffering from enormous shifts caused by new data, this book, as well as its accompanying NOVA/PBS video presentation, still sets the benchmark against which all such efforts must be measured.

Smolin, Lee. *Three Roads to Quantum Gravity.* New York: Basic Books, 2001. This book offers a lively debate about alternatives to superstring/M-theory. In the process, the author clearly enunciates the issues to be resolved by a mythical quantum theory of gravity. This book should provide some thought-provoking considerations for those asking whether Einstein's dream could be realized by some work other than superstring/M-theory.

Suplee, Curt. *Physics in the 20th Century.* New York: Harry N. Abrams, 1999. The highlight of this book is its visual offerings taken from diverse realms of the field of physics. It provides a balanced view of the various strains of physics from the 20th century. In particular, it covers background material to much of the discussion of the quantum world and the geometry of relativity in this course.

Susskind. Leonard. *The Cosmic Landscape: String Theory and the Illusion of Intelligent Design.* New York: Little, Brown, 2005. One of the fathers of the field of string theory tackles some of the thorniest current issues in the field. About 20 years ago, it was the predominant view that the solution to the equations of string theory would be unique and could describe our universe completely. Currently, a substantial number of physicists believe the opposite.

Superstring/ M-theory may possess a large, perhaps infinite, number of consistent solutions that today are sometimes said to be elements of a "landscape." Implications of this idea for our existence are considered in this challenging read.

Tyson, Neil De Grasse, and Donald Goldsmith. *Origins: Fourteen Billion Years of Cosmic Evolution*. New York: W.W. Norton, 2004. If Sagan's *Cosmos* has a successor in tackling such an enormous wealth of issues in thinking about the cosmos and humanity's place in it, this book is as likely a contender for this title as any I know.

Walpole, Horace. *Horace Walpole and His World: Selected Passages from His Letters*. Boston: Elibron Classics, 2005. The word *serendipity*, in fact, had a beginning that was…serendipitous. This book chronicles some of the life and ideas of the man who brought this word to prominence in the English language.

Weinberg, Steven. *The First Three Minutes*. New York: Basic Books, 1993. Though this book is dated, it is still one of the finest expositions of what modern physics can say about the evolution of the universe near the time of its beginning. This work was probably the first to open for the public a presentation of what science can say on these issues.

Wells, H. G. *The Time Machine*. New York: New Review, 1985. The concept that time is the fourth dimension did not originate with Einstein. When he was 16, there was already a quite well known science fiction novel by H. G. Wells that possesses a surprisingly lucid description of this idea, which was already several millennia old. However, Einstein's 1905 work was the first to point out exactly what was wrong with Wells's ideas (and all who came before).

Wigner, Eugene. "On the Unreasonable Effectiveness of Mathematics in the Natural Sciences," in *Communications in Pure and Applied Mathematics*, 13(1). This essay clearly speaks to the unexpected magnitude of success that the mathematical paradigm set out for the field of physics by Galileo and Newton.

Zee, A. *Fearful Symmetry: The Search for Beauty in Modern Physics*. Princeton, NJ: Princeton University Press, 1999. This book is still one of the best available to the public to clarify the power of the concept of symmetry in the advance of fundamental physics since the time of Einstein.